Forest Fires and Their Control

Other Related Books

Agroforestry Theory and Practices	*Antony Joseph Raj*
Forest Tribology and Anthropology	*Mhaiske V.M.*
Forestry - A Subjective Guide for IFS Aspirants	*Parthiban K.T.*
Forestry Principles and Applications	*Raj A.J.*
Forestry Research Extension: Challenges & Strategies	Dange C.S.
Forestry Technologies A Complete Value Chain Approach	*Parthiban K.T.*
Industrial Agroforestry : Perspectives and Prospectives	*Parthiban, K.T.*
Multipurpose Plants	*Srivastava, P.K.*
Nursery and Plantation Practices in Forestry	*Kumar V.*
Objective Forestry : for All Competitive Examination, 2nd Ed.	*Parthiban K.T.*
Pests of Forest Importance & their Management	*Tyagi B.K.*
Plant Nursery Management : How to Start and Operate A Plant Nursery	*Ray P.K.*
Plantation and Agroforestry Pulpwood Value Chain Approach	*Parthiban K.T.*
Practical Manual on Plantation Forestry	*Panwar P.*
Silvipasture in India: Present Perspectives and Challenges Ahead	*Palsaniya D.R.*
The Identification of Trees and Shrubs 2nd Ed.	*Makins F.K.*
Tree Moratality Assessment and Mitigation	*Negi J.D.S.*
Tropical Forest Ecosystems Structure and Function	*Singh, V.P.*

Forest Fires and Their Control

E.S. Artsybashev

United Book Prints

First reprint in India, 2017

Published by:
United Book Prints
4806/24, Ansari Road,
Daryaganj,
New Delhi - 110 002
Tel.: 011-41511055
E-mail: unitedbookprints@gmail.com

Print: 2017

ISBN: 978-93-83692-59-0

Printed in India

UDC 634.0.43

The nature of forest fires, their development and spread depending on the conditions of forest vegetation and a combination of various meteorological factors has been described on the basis of scientifically collected data. The author describes the designs of various fire observation towers and masts, and proposes unique technical methods ensuring the safe ascent of the observer to the top as well as his descent with the minimum expenditure of manpower and also describes technical equipment and methods of detecting and extinguishing forest fires. While examining the prospect of using television and the infrared technique for detecting fires, the author also discusses in detail aerial suppression of forest fires (using airplanes and helicopters), including the method of controlling extensive fires by means of artificially induced rainfall.

Contents

Introduction

Forest fires are a calamity for forests of temperate climate zones in view of the fact that the damage they cause to the forest economy is considerably greater than all the damage caused by harmful insects and diseases of wood taken together. Forest fires retard the vital activity of forest crops, subsequently encouraging the multiplication of pests and fungal diseases. Fire destroys not only greenwood and harvested timber, but also construction material in the forest, many species of animals and edible birds as well as fodder grown for domestic animals, etc. Forest fires destroy the raw material resources of the timber industry, causing untimely closure of industrial bases or their transfer to other regions. In case of large fires which acquire the character of a natural calamity, the extent of atmospheric smoke content can be so high that water and air transport come to a standstill for long periods.

In spite of the development of many fire control methods and techniques, large forest fires continue to remain a common phenomenon. Therefore, the problem of controlling forest fires has been and remains a most urgent issue in the forest economy of our country, for without a well-organized forest conservation service, all other forestry improvement measures are meaningless.

The forests of the Soviet Union consist mainly of coniferous trees (pine, larch, fir, spruce, cedar), which are highly susceptible to fire. The absence of any forest conservation service in prerevolutionary Russia resulted in forest fires, affecting millions of hectares. Thus, in 1915, an area of nearly fourteen million hectares of forest in western Siberia was damaged by fires, which exceeds the total forest area of France (Antsyshkin, 1957). The smoke cloud over Siberia occupied an area equal to the whole of Western Europe, delaying crop maturation for a period of three weeks. In 1921, when a severe drought struck the Volga bank region, the forests of the Mari Autonomous Soviet Socialist Republic continued burning throughout the summer. Even though over thirty-five thousand people were engaged in fighting the fury of the fire for several days, nearly three hundred thousand hectares of mature commercial wood were destroyed.

During the years of the early five-year plans in the period of vigorous development of the national economy, the situation in regard to forest protection underwent an abrupt change with the setting up of a State Forest Conservation Service. The Aerial Protection Service was organized to assist in the conservation of the remote and thickly forested regions of the North, Siberia and the Far East. With the adoption of new methods of controlling forest fires including the organization of fire-retardant chemical stations, and the use of the latest machines and means of transport, especially aircraft, a marked improvement in the mobility and efficiency of forest fire brigades was noted. As a result of the measures taken by the Soviet Government, forest fire risks were reduced to less than half compared with the initial postwar years, the average area of each fire being reduced in the same proportion.

According to the data collected over several years, nearly 98 percent of forest fires have been liquidated over areas not exceeding 200 hectares, which indicates the high efficiency of the forest conservation service. The area of forests covered by these fires does not exceed 20 percent of the total affected area. The remaining 80 percent, affected by large fires covers hundreds of thousands of hectares, although the number of such fires does not exceed one and a half percent of the total number (Chervonnyi, 1973). Such fires are most often the result of mass conflagration (e.g. due to lightning), after prolonged drought, accompanied by strong winds. These are difficult to extinguish by the usual methods. Therefore, the most urgent task is the further development of the means and methods of prevention, suppression, and extinction of such fires which fall in the category of natural calamities.

With intensification of forest management, the risk of forest fires will steadily increase due to the accumulation of deadwood resulting from the use of arboricides in forest management, the drainage of marshes and bogged up forest stands, and an increase in timber fellings and of areas under conifers. Commercial development of the forest tracts of Siberia and the Far East region of the USSR, development of tourism and a private transport system will also add to the causes of fire hazards. This increase in fire hazard should be countered by a well-organized ground and aerial forest conservation service, equipped with the latest and most effective means of prevention, detection and control of forest fires.

Forests are to be protected from fire primarily in order to utilize them. Therefore, between timber felling and forest conservation there can be no contradiction. Unlike other natural resources (oil, coal, mineral ores, etc.), forests have the capacity of renewal. However, one must be concerned as to the proper time to begin felling in the forest in order to harvest timber—whether it should be now or after 150-200 years. Large areas laid bare by fires in the Enisei and Angara pine forests, and in Baikal forests and Khabarovsk territory should stand as a warning that timber as a resource is not inexhaustible. The commercial timber that is burnt and the trail saplings of timber species raised through seedlings in the same area do not give identical results. Timber grown on burnt areas can only be utilized by the next generation. If afforestation is expected to replace bare rock, then the harvesting time may be delayed by another crop.

The problem of forest fire is complex and varied, there being two main resultant aspects, viz., the social aspect which involves instilling a protective attitude toward the forest among the workers, and the technical aspect related to the development of means and methods of detection and liquidation of forest fires.

In recent times, the question of protection of forests against fire has attracted great attention. Requirements of conservation of natural resources, including timber, were clearly laid down in the directives of the XXIV Congress of the Communist Party of the Soviet Union (CPSU) in the resolutions of the Council of Ministers of the USSR, dated 10th January, 1973, which require the following: "Strengthening of the means of conservation and improvement of the utilization of natural resources". Forestry institutions and social organizations have been conducting extensive work on explaining these important resolutions to the broad masses of workers.

Thus, protection of forests against fire has become one of the most important tasks of the Government. However, such an objective cannot be achieved without the State Forest Conservation Service using technical equipment and effective means and methods of detecting and liquidating forest fires, based on the latest achievements of science and technology. In this regard, extensive work is being done by field scientific research institutions, such as the Leningrad Research Institute of Forestry and the Far-Eastern Scientific Research Institute of Forestry, as well as the experimental production laboratory of the Central Air Force base. Significant contribution to the theory and application of forest fire control has been made by the Wood Pyrology laboratory of the Forest and Timber Institute of the Siberian branch of the Soviet Academy of Sciences, and the Arkhangel'sk Institute of Forest and Wood Chemistry. The organizations taking part in solving individual problems of an allied nature are: the Design Bureau of the Ministry of Aviation Industry and the Ministry of Road and Highway Machine Building, the Research Institute of the Ministries of Civil Aviation, Light and Chemical Industry and the Main Administration of the Hydrometeorological Service. As a result of their work, substantial experience has been gained on development of the techniques of detection and liquidation of forest fires, which has considerably enriched the science of wood pyrology in the country. In its turn, the technique of forest fire control has been developing on the basis of recent knowledge about the nature of origin and development of forest fires and their interrelationship with forest vegetation and meteorological factors.

Nature of Forest Fires

GENERAL INFORMATION ON COMBUSTION

Man has traversed a very long way in his understanding of the combustion process in forests. Even at the dawn of evolution, primitive man observed how lightning can reduce a tree to ashes, although he failed to understand this phenomenon. With existing developments in the fields of chemistry and physics, the mechanism of combustion is basically understood, even though some of its aspects, especially those concerning the burning of forest combustibles, still require further investigation.

Combustion is generally conceived as the rapid chemical reaction combining a combustible material with oxygen, accompanied by a release of heat and a luminous radiation. From the chemical point of view, combustion is nothing but an oxidation reaction accompanied by the formation of new matter and the liberation of excess heat accumulated in the molecules of combustible matter.

According to the theory advanced in 1898, heating of the combustible system (matter + oxygen) activates oxygen by rupturing one interatomic bond. The molecule of oxygen with one free valency bond then combines with the combustible matter, forming a peroxidate of the type $R_1–O–O–R_2$ or R–O–O–OH. In chemistry, this theory has acquired the term 'peroxidation' (Alekseev and others, 1971). However, it cannot explain many phenomena which are unexpected at first sight, but are known to occur during oxidation, for example, the catalytic phenomenon, occurring with the addition of a small quantity of another substance having a definite composition and concentration which sharply accelerates the oxidation reaction. Also, this theory cannot explain the phenomenon of spontaneous combustion which is often encountered in practice.

The mechanism of the combustion process was lucidly explained by Academician N.N. Semenov (1957) who worked out the structural theory of chain reactions. According to his theory, the process of combustion (oxidation) begins with activation of combustible matter, accompanied by a gradual rise in temperature. Under certain conditions, this process begins to accelerate automatically, resulting in spontaneous ignition of the substance.

The autoignition point is different for various substances and depends on the concentration of the substance, its density, porosity, humidity, pressure, etc. For example, the autoignition point of yellow phosphorus in the open air does not exceed 15°C, while that of milled peat is almost

400°C. Autoignition of peat is the result of biological and chemical processes which occur at a definite humidity and relatively low temperature.

Apart from peat, hay, sawdust, forest floor, leaves, grass, etc., collected in large heaps and having a definite humidity, also possess the propensity of autoignition. However, in the practice of forest fire control, the process of ignition (kindling) is encountered more often. Ignition differs from autoignition by the fact that in the first case, a part of the combustible substance is heated by a heat source applied from outside, whereas in the second case, the rise in temperature occurs due to the internal energy of the combustible substance itself, and simultaneously in its entire mass. It is important to understand this when examining the causes of forest fires.

Oxidation processes may take place at extremely varying rates—from very slow to a burst, occurring within a thousandth part of a second. An example of the former is decomposition, i.e. the oxidation process of organic matter occurring due to the activity of bacteria, while that of the latter is the oxidation reaction occurring in mercuric fulminate. If combustion is accompanied by liberation of fuel gases, a flame may be observed, for example, during the combustion of wood.

Combustion, as encountered in daily life, is an extremely complicated process. The combustion of a candle and that of belt of forest fire are extremely different from each other, although in both cases, the sequence of

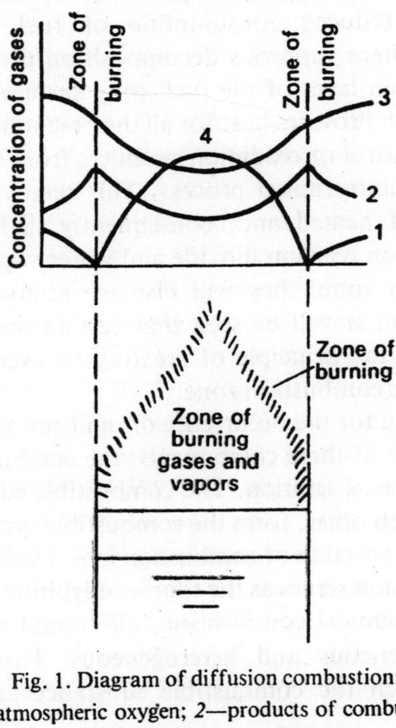

Fig. 1. Diagram of diffusion combustion:
1—atmospheric oxygen; *2*—products of combustion;
3—fuel vapors; and *4*—fuel gases.

the combustion process is similar. To simplify the comparison, fuel combustion in a wick burner will be examined as a model of diffusion combustion (Fig. 1). Liquid fuel is absorbed by the wick and as a result of capillary action rises to its upper part where, under the effect of the heat of the flame, it vaporizes, and under the action of natural draft it rises to the combustion zone. A luminescent zone, consisting of incandescent particles of carbon and fuel transformed into a gaseous state, is observed in the upper part of the fuel spray. Traces of soot can be seen on a glass rod momentarily introduced into the flame in the zone at the very center of the spray. The combustion zone is the thin shining layer on the surface of the spray where vaporous fuel enters from within, and oxygen from the surrounding air. The process of combustion of a particular layer of fuel mixture lasts a fraction of a second but continues uninterrupted due to the entry of new portions of the fuel gas. In the combustion zone at the edge of the spray, the concentration of oxygen and the fuel is zero (the points of intersection of curves 1 and 4), although the maximum concentration of combustion products is observed (curve 2). In the vaporous and gaseous zones (curves 3 and 4) combustion is absent since all the oxygen in the combustion zone has entered into the reaction.

As seen from the diagram, a number of preparatory stages precede the process of combustion: it begins from the preheating of the fuel and ends with its transformation into the gaseous state. The preparatory stages, as compared to the combustion process, proceed very slowly, with some time lag, thus ensuring reduced consumption of fuel. The zone of true combustion itself, where vaporous decomposition products combine with oxygen, occupies a thin layer of the fuel spray surface. However, it is this 'boundary layer' which provides heat for all the preparatory stages.

A continuous removal of oxidation products from the combustion zone is essential for an uninterrupted process. This removal is ensured by the incoming currents of heated and, consequently, lighter air. If the end products of combustion (carbon dioxide and water vapor) are not removed from the combustion zone, they will displace atmospheric oxygen, and combustion will cease. It will be seen that certain methods of forest fire control are based on the principle of creating an excess concentration of these substances in the combustion zone.

The basic condition for the occurrence of, and sustaining the combustion process is the presence of three components: the combustible substance, the oxidizer and the source of ignition. The combustible substance and oxygen, which interact with each other, form the combustible system, while the source of ignition induces the reaction of combustion in it. Under steady-state conditions, the zone of reaction serves as the source of ignition.

Based on their chemical composition, all combustible systems can be divided into homogeneous and heterogeneous. Homogeneous systems include those in which the combustible substance and the oxidizer are uniformly mixed with each other (for example, a mixture of gasoline vapors and air). The combustion of uniform systems is called *homogeneous combustion*. In combustion of nonuniform chemical systems, the substance and the oxidizer are not mixed with each other, but have an interface (for example, the combustion of wood, coal, oil etc.). The combustion of nonuniform systems is called *heterogeneous combustion*. The nature of homogeneous combustion can best be explained by the example of the ignition (oxidation) of methane. A molecule of methane (CH_4) has one atom of carbon with all its four covalent bonds occupied by hydrogen atoms. (It should be remembered that the calorific value of one kilogram of hydrogen is equal to 34180 kcal/kg, while that of carbon is 7840 kcl/kg). Of all the known hydrocarbons, the hydrogen content in methane is the highest (25%). Therefore, methane has a very high calorific value (13300 kcal/kg). If oxidation of methane is affected gradually, it is possible to obtain a number of new organic substances with lowered calorific value since a part of the energy is used up during the reaction itself. This is illustrated by the combination of one oxygen atom with a molecule of methane which gives methyl alcohol, with a calorific value of only 5340 kcal/kg. Addition of one more oxygen atom gives a new product, namely formaldehyde, with an even lower calorific value (2790 kcal/kg) and water. Addition of the third atom leads to the formation of formic acid, the calorific value of which is even lower (1360 kcal/kg). Finally, addition of the fourth atom results in the formation of carbon dioxide and water, the calorific value of which is equal to zero. The last reaction is the final one, when thermal energy reserves of this gas are used up. Combustion (oxidation) of methane, as of any fuel mixture, can be effected only with a definite concentration of oxygen or air. With excess or deficiency of these oxidizers, the mixture will not burn.

The volume of air V_a required for combustion of one kilogram of a combustible substance of a complex chemical composition as, for example, hydrocarbons, can be determined if the weight percentages of carbon, C^p, hydrogen, H^p, and oxygen O^p present in its composition are known. Calculation is done according to the formula

$$V_a = 0.89\ C^p + 0.265\ H^p - 0.0333\ O^p\ (m^3/kg).$$

Determination of the volume of air for combustion of a fuel mass unit is not only of theoretical interest, but can also have practical value in working out methods of extinguishing fires based on displacement of atmospheric oxygen from the combustion zone. Combustion of one kilogram of wood having 7% humidity, requires 4.18 m^3 or 5.9 kg of air. During combustion, especially of the diffusion type, the air flow rate is actually 2-7 times more than that calculated theoretically.

The main characteristic of the heterogeneous process is the capacity of a substance to burn with a small concentration of atmospheric oxygen. An obvious example of this can be charcoal in a covered brazier or in deep layers of peat. It will be seen later on that this peculiarity of heterogeneous combustion plays an important role in forest fires. Heterogeneous combustion, for example, that of peat, cannot occur without access to atmospheric oxygen, because of the oxygen in the tissues of the combustible substance itself being in the bound state. This aspect can be easily proved by conducting the following simple experiment. Place a small piece of burning peat in a test tube and cover it with a tight stopper. In the beginning there will be almost no slackening in combustion, since it takes place due to the fact that some oxygen from the air remains inside the test tube, with the intensity of combustion gradually decreasing and stopping completely after 1.5 to 2 minutes. Since the bound molecular oxygen in the peat mass (about 34%) cannot sustain combustion, a flow of oxygen from outside is essential. Deficiency of atmospheric oxygen in the combustion zone in deep-seated peat fires is only of the reasons for their slow spread.

As the preheating of the combustible material becomes more intense, the optimum concentration of fuel gases at which the mixture can burn is arrived at more rapidly at a lower temperature. This rule is confirmed by the experiments of V.P. Molchanov (1957) in which pine cones caught fire on intense heating at a temperature of about 200°C, whereas with slow preheating, the ignition temperature of the cones increased as much as 400°C. If preheating is still slower, the optimum concentration of fuel gases in the air may not be reached, and instead of the flame phase of combustion in the cones, carbonization takes place.

COMBUSTION OF WOOD

As is well known, wood is composed mainly of cellulose ($C_6H_{10}O_5$) and lignin, an organic compound the formula of which has not been established so far. The composition of wood includes a large quantity of resins and essential oils, which have a comparatively high calorific value. Oxygen constitutes nearly 42% of the total wood content, which makes wood a comparatively oxidized substance. Therefore, it cannot liberate a large quantity of heat during its final oxidation in the process of combustion, as for example, coal or oil. However, preliminary oxidation makes the hydrocarbon molecule highly unstable. Such a molecule can decompose at low preheating temperatures. In the process of decomposition, gaseous substances are partly liberated ('volatile matter') which, on mixing with air, begin to burn with a flame. That is why wood is one of the comparatively easily ignitable materials.

Figure 2 shows a diagram of the combustion process of wood from the preheating to the coal phase. Depending on the rise of temperature, the following stages of this process may be established:

I—from 0 to 100°C—the process of heating of the particle due to combustion of adjoining particles;

II—from 100 to 150°C—evaporation of moisture (free, capillary, absorbed);

III—from 150 to 300°C—growing process of decomposition of wood with liberation of fuel gases and advent of carbonization;

IV—from 300 to 500°C—gradual slowing of the process of decomposition within the carbonized particle; and

V—from 500 to 1000°C—combustion of coal with liberation of carbon monoxide and carbon dioxide.

In direct proportion to the removal of volatile matter from the particle, the luminous flame gradually begins to decrease and, finally, disappears completely. The carbon particle surrounded by a semitransparent bluish flame of atmospheric carbon monoxide remains. As a result of intense oxidation with atmospheric oxygen, the carbon particle burns up quickly and is transformed into ash.

While examining the diagram of combustion, one may agree with the observation of G.F. Knorre (1959) that wood does not burn at the surface, but decomposes from the surface. Actually, only fuel gas burns when it is distinguishably separated from the wood surface, since the flame extends

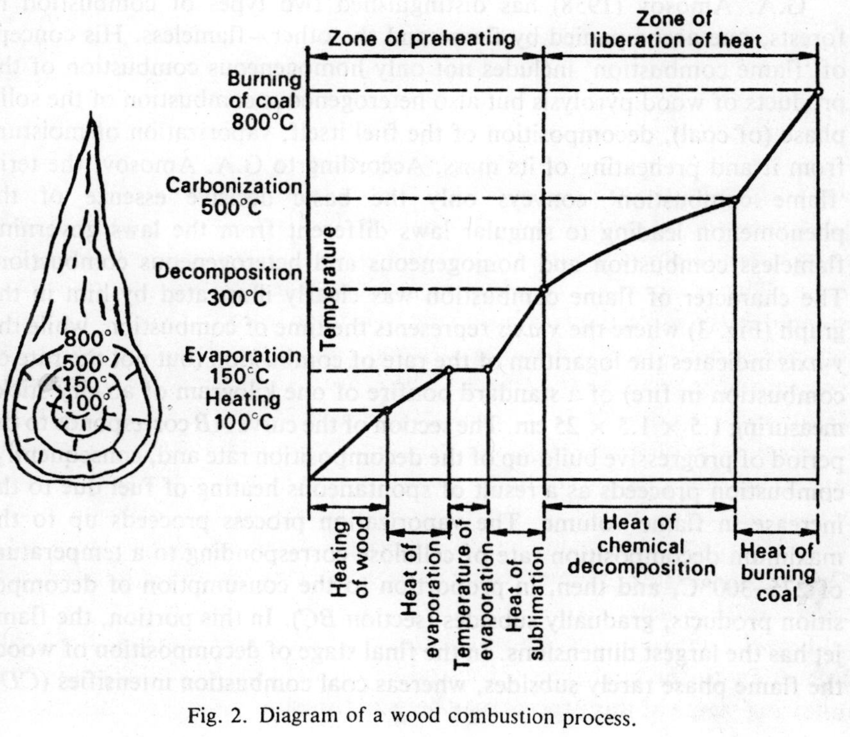

Fig. 2. Diagram of a wood combustion process.

upward due to the natural thrust of air. The emerging fuel forces back the air from the surface of the heated wood and intercepts the atmospheric oxygen before it reaches the wood surface. In this case, we have an example of homogeneous diffusion combustion.

Sometimes, a part of atmospheric oxygen reaches the decomposition zone of wood, sharply intensifying the process of its gasification. Such a phenomenon may be observed when wind sharply tilts the flame, exposing the coal surface of wood to combustion. But since the atmospheric oxygen in this zone is not sufficient for complete oxidation of the sublimates, carbon monoxide (CO) is formed there which, on contact with atmospheric oxygen, itself burns forming carbon dioxide (CO_2). What occurs here is heterogeneous combustion, i.e. combustion with 'oxygen deficiency'. Thus, combustion of wood is a highly complex physicochemical process in which the characteristics of homogeneous as well as heterogeneous combustion are inherent.

Since wood contains a substantial quantity of its own oxygen the flame produced is not very bright. At the time of oxygen deficiency all the particles of solid carbon do not burn in a flame and a portion of these gets carried upward by a convection current, forming smoke. Apart from carbon particles, smoke contains vaporous moisture and unburnt gases. The darker the smoke, the larger the content of carbon particles in it.

G.A. Amosov (1958) has distinguished two types of combustion in forests, one accompanied by flame and the other—flameless. His concept of 'flame combustion' includes not only homogeneous combustion of the products of wood pyrolysis but also heterogeneous combustion of the solid phase (of coal), decomposition of the fuel itself, vaporization of moisture from it and preheating of its mass. According to G.A. Amosov, the term 'flame combustion' conveys only the basic decisive essence of the phenomenon leading to singular laws different from the laws governing flameless combustion and homogeneous and heterogeneous combustion. The character of flame combustion was clearly illustrated by him in the graph (Fig. 3) where the *x-axis* represents the time of combustion, while the *y-axis* indicates the logarithm of the rate of combustion (but not the rate of combustion in fire) of a standard bonfire of one kilogram of air-dry sticks measuring 1.5 × 1.5 × 25 cm. The section of the curve *AB* corresponds to the period of progressive build-up of the decomposition rate and, consequently, combustion proceeds as a result of spontaneous heating of fuel due to the increase in flame volume. The vaporization process proceeds up to the maximum decomposition rate of cellulose corresponding to a temperature of 275–300°C, and then, in proportion to the consumption of decomposition products, gradually subsides (section *BC*). In this portion, the flame jet has the largest dimensions. In the final stage of decomposition of wood, the flame phase rarely subsides, whereas coal combustion intensifies (*CD*).

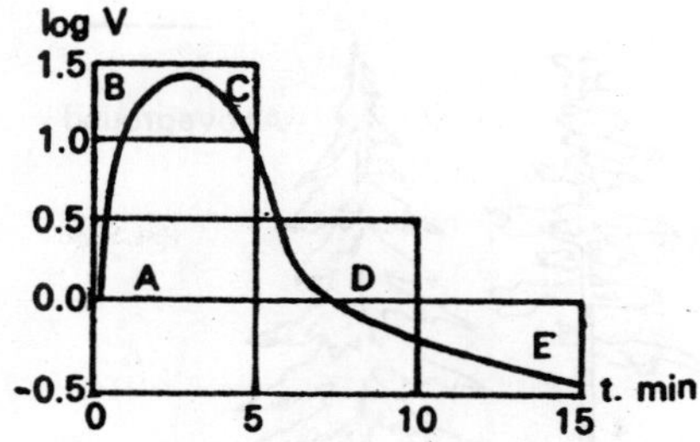

Fig. 3. Graph of the logarithm of the flame combustion rate of wood against time.

The portion *DE* mainly characterizes the flameless phase, i.e. the phase of combustion of carbonic residues before transformation into ash. During the period of flame combustion, up to 70% of the total heat is liberated, the remaining 30% is imparted by the coal phase of combustion.

CHARACTERISTICS OF FOREST COMBUSTIBLES

The elements of combustion in a forest constitute a biogeocenotic complex having a three-dimensional structure and distinguished by a great variety of components. The main components are: the forest floor, lichens, mosses, top humus, stubs and dead wood, bushes and grasses, peat, undergrowth and seedlings, coniferous needles, leaves, twigs, branches in the tree canopy and, finally, tree trunks. Knowledge of their main characteristics (structures, composition, humidity and calorific value) is very important for understanding the process of occurrence and development of forest fires, as well as for working out preventive measures, means and methods of their control.

A number of works have been devoted to the study of forest combustibles. Data on this question given in the works of I.S. Melekhov (1939, 1947), A.A. Molchanov (1949, 1952, 1954), V.G. Nesterov (1945) and A.A. Korchagin (1954). Extensive investigations in this direction were conducted by S.M. Vonskii (1957) and V.P. Molchanov (1965), scientists at the *LenNIILKh**. In 1970, N.P. Kurbatskii gave a detailed analysis of these works, and, by adding his own observations, attempted to classify various components of forest biogeocenosis, basing this classification on their pyrological nature.

Considering the canopy levels and morphological structure, combustibles can be divided into three main groups (Fig. 4), in relation to the character of forest fires: I—surface, II—aboveground and III—under-

**LenNIILKh*—Leningradskii nauchno-issledovatel'skii institut lesnogo khozyaistva (Leningrad Scientific Research Institute of Forestry)—Translator.

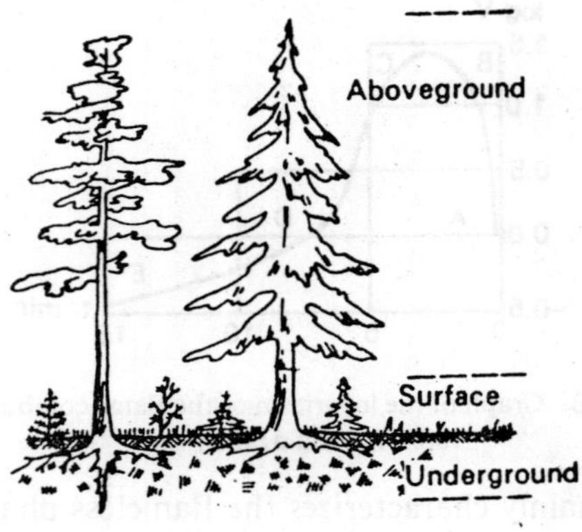

Fig. 4. Stagewise distribution of different groups of combustible materials in a forest.

ground. The first group includes all the organic mass of the cover present on the surface and coming in close contact with it (mosses, lichens, grasses, bushes, top humus, forest floor), i.e. everything which usually burns up in a ground fire. This group of materials should also include undergrowth representing a combination of bushy and sometimes woody trees that can never reach the upper story of the forest stand and therefore never replace it. The first group of combustibles is characterized by great porosity and hygroscopicity which makes it react quickly to all changes in weather conditions.

The second group (surface combustibles) includes all trees, comprising the upper story, and undergrowth. This group includes tree trunk along with lichens on them, twigs, branches, needles and foliage. It is characterized by a comparative independence among individual components, low hygroscopicity and, consequently, negligible change in humidity under the effect of weather conditions.

And, finally, the third group of combustibles (underground) includes all the components of organic origin situated below the ground surface level (peat, organic composition of soil and the underground part of all types of vegetation), i.e. the root systems of grasses, bushes and trees. These components have a comparatively compact structure, and their combustion most often occurs in a flameless phase (smoldering).

The main components of the surface group of forest combustibles include live soil cover and top humus. N.P. Kurbatskii (1970) calls them conductors of combustion. Under favorable conditions, they can burn with a liberation of a quantity of heat which is more than sufficient for sustaining and propagating the process of combustion. This primary combustible group is the primary combustible matter, the ignition of which starts almost all fires.

In lichen-moss-covered, moss-covered and ledum-covered pine forests which are most inflammable in the northwestern regions of the USSR, the live soil cover is mainly represented by bushes of various types (heather, maintain cranberry, bilberry and ledum), bushy lichens (*Cladonia alpestris, C. rangiferina, C. silvatica*) and green mosses dominated by *Pleurozium schreberi*. In modern forest typology, the living soil cover determines the characteristics of the forest types. Top humus includes small undecayed morphological parts of trees and bushes: needles, leafage, bark, cones and twigs.

In the types of forests mentioned above, no great difference is observed up to the 5-year old crops in the total soil cover reserves (live cover + top humus) in an absolute and dry state. In lichen-moss and ledum-covered pine forests above the 5-year old crops, the soil cover contents are more as compared to moss-covered pine forests (Vonskii, 1957). In dense lichen-moss pine forests, the weight of soil cover varied from 0.8 kg/m^2 in the 2-year age group up to 1.4 kg/m^2 in the 7-year old crops; in moss-covered pine forests, it varied from 0.7 to 0.9 kg/m^2 in forests of the same age groups, while in ledum-covered forests, the weight varied from 0.9 to 1.4 kg/m^2 for the 4–7 year old crops. The representative character of various components of soil cover in the studied forest types varies in a percentage ratio by mean values (Table 1). The data obtained indicates that the main combustibles in the soil cover are mosses, lichens and top humus. It will be seen that these are the first to ignite in a forest due to various fire sources and that they determine the intensity of ground fires.

TABLE 1. REPRESENTATIVE CHARACTER OF SOIL COVER COMPONENTS IN VARIOUS TYPES OF PINE FORESTS (ACCORDING TO VONSKII, 1957)

Forest type	Components of cover, in volume (wt.%)		
	bushes	mosses and lichen	top humus
Lichen-moss-covered	6–18	38–62	32–54
Moss-covered	0–9	19–67	20–81
Ledum-covered	28–58	24–43	10–31

Detailed data on soil cover contents in the most common types of forest in the northwestern part of the USSR, Western Siberia and the Central Siberian plateau are given in the work of N.P. Kurbatskii (1970). In lichen-covered pine forests (Leningrad Province), the weight of soil cover contents varied: from 0.89 kg/m^2 in 4-year old crops to 1.43 kg/m^2 in 7-year old crops; in cranberry-covered pine forests, these figures were: from 0.86 kg/m^2 in 5-year old crops to up to 1.27 kg/m^2 in 2-year old and up to 1.15 kg/m^2 in 3-year old crop. This data confirms the quantitative character of soil cover, given earlier by S.M. Vonskii (1957) for the most inflammable types of forest in the northwestern regions of the USSR.

For more humid types of forest, the following weight characteristics of soil cover were obtained: for sphagnum-ledum pine forests of the 6-year old crop it was 1.27 kg/m^2; bushes and grasses proportionately constituting nearly 34% of the total cover weight; for sedge sphagnum pine forests of the 5-6 year old crops—from 0.83 to 1.44 kg/m^2 (the proportion of bushes from 43 to 53%); for moss-covered spruce forests of the 3-year old crop—from 1.04 to 1.15 kg/m^2 (grasses and bushes were absent). In ledum-covered spruce forest of the same age, but of a more complex composition (with an admixture of pine and the 2nd story consisting of spruce), the cover reserves did not exceed 0.84 kg/m^2, and only in the haircap-moss spruce forest of the 4-year old crop with an admixture of pine, it reached an amount of 1.7 kg/m^2, which is a record for this region. For typical forest stands of Western Siberia and the Krasnoyarsk region, the soil cover contents vary within a wide range but do not exceed 2 kg/m^2.

The results obtained confirm the conclusion of A.A. Molchanov (1960) to the effect that the total quantity of soil combustibles is much higher in forest types with excess soil humidity, where decomposition of organic matter is slower. G.V. Snytkin (1969), who studied the composition of combustibles in the larch forests of Western and Eastern Siberia, arrived at the same conclusion. His observations show that, on the whole, the soil cover composition in these forests increases as one moves eastward. This, according to him, is explained by the microclimate of the northern and eastern regions which is unfavorable for decomposition of plant residues. However, as compared to other exposures, the microclimate of the northern slopes is more favorable for the accumulation of combustibles. G.V. Snytkin also established that combustible contents per unit area depend on their situation on the slope. According to his data, in the lower parts of the northern slopes under conditions prevalent in the Magadansk Province, the soil cover reserves constitute 2.55 kg/m^2; in the middle part, it is 2.18 kg/m^2, while in the upper part, it is only 0.82 kg/m^2. This means that as one approaches the watershed, these reserves are markedly reduced.

The quantity of surface combustibles in the same types of forest during the fire risk season does not remain constant. It undergoes drastic changes in forests dominated by herbaceous plants and lesser changes in forests with moss and evergreen bush cover. In early spring, the grassy cover contains small reserves of dry grass resulting from preceding growth periods which encourage the spread of rapid ground fires. In summer, the contents of the green mass of the grassy cover sharply increases, on the whole reducing fire risk in the forests. During autumn, in proportion to a general drying up of the cover, the fire risk in the grassy type of forest, again increases. In other types of forests, the soil cover contents depends mainly on the period that has elapsed after the last fire. Periodically recurring ground fires in the dry woods of the north, and in Siberia seem to regulate the total soil cover contents, and thus considerably reduce combustion intensity in recurrent fires.

The most complete account of the role of grasses and bushes according to forest types and the time of the year, in forest fires was given by I.S. Melekhov (1947). In the spring dry grasses increase the risk of fire in forests, while in summer green reed grass retards the spread of fire. Coarse herbs consisting of spirea, monkshood, fireweed and ferns reduce the possibility of forest fires spreading. On the other hand, during very dry weather, marsh tea, together with cassandra, bog whortleberry, bog rosemary, and andromeda (moorwort) intensify the ground fire flame until the tree crowns are affected. Heather, which develops into dense thickets, is highly inflammable. A number of investigators (A.A. Korchagin, 1954; I.N. Balbyshev, 1949) have noted the fire-resistant quality of manzanita, stone clover, pyrola, club mosses, bergenia, perennial crowfoot, Sakhalin buck-wheat and other plants. M.A. Sofronov (1967) has observed in his paper devoted to the study of fires in mountain forests of southern Siberia that the fire risk season in these regions is over when the grass contents are reduced to ½ to ⅔ of the total composition.

Lichens and some mosses have a very highly active surface which determines their hygroscopicity, humidity and inflammability. Lichens have the most active surface, Shreber's moss has a less active surface, followed by bog sphagnum and the least active among these is of haircap moss. Their inflammability has also been established in the same sequence. The easiest to ignite are lichens, while haircap moss may be classified as 'fireproofing agent', i.e. fire-resistant moss. The latter property of this type of moss is also explained by the fact that its ash content is higher than other mosses mentioned above. On the other hand, the tissues of lichens contain a high percentage of albuminous substances and very little ash.

The most important factor determining the inflammability of any soil cover component is its humidity. In published works, voluminous data is available on this subject, but these are difficult to compare, since measurements by different authors (P.P. Serebrennikov and V.V. Matreninskii, 1937; I.S. Melekhov, 1939; V.G. Nesterov, 1954; A.A. Korchagin, 1954; S.M. Vonskii, 1957; V.A. Zhdanko, 1965; N.P. Kurbatskii, 1970; M.A. Sofronov, 1970, V.I. Zhukovskaya, 1970 and others) were carried out at different times under different conditions and by different methods. In order to somehow systematize the data, it is proposed to establish three 'humidity levels' for each type of combustible: the minimum, usually confined to the end of the droughty period; the maximum corresponding to the maximum water-retaining capacity of the combustible after prolonged rainy weather, and the critical, corresponding to the maximum humidity under which combustion of the component is possible. For the application of forest fire control, the last category is more important than the other two.

P.P. Serebrennikov and V.V. Matreninskii (1937) have established the lower humidity levels: for lichens 6–8%, for Shreber's moss 18–22%, for haircap moss 28–32% and for sphagnum 47–57%. I.S. Melekhov (1939) observed that on cutting lichens six days after rain, the humidity dropped to 13.3%, in green mosses to 20% and in haircap moss to 44%. A.A. Korchagin (1954) demonstrated that bushy lichens absorb 2.5 times more water as compared to their air-dry weight. Therefore, the maximum humidity level of this type of combustible will be nearly 250%. For green mosses, the maximum level, according to his data, is still higher, viz. 500–700%, while for sphagnum, it is 1500–3100%. According to the data of V.I. Rutkovskii (1936), the soil cover of lichens can retain up to 4.5 mm of rainfall while the capacity of green mosses is up to 10 mm (Kurbatskii, 1970).

The critical humidity of combustibles, under which the spread of fire over the soil cover becomes possible, is of great practical interest in determining fire risk. According to the data of V.G. Nesterov (1949), the critical humidity level for a cover consisting of lichens, greenmosses and haircap moss corresponds to 70–80%. In case of absence of live soil cover, the inflammability of dead soil cover in pine forests was completely lost on attaining 47% humidity. However, most fire cases (48 out of 50) were registered with soil cover having humidity up to 30%. Detailed investigations in this direction were conducted by S.M. Vonskii (1957) for the northwestern regions of the European part of the USSR. Table 2 shows that the maximum values of humidity of soil cover combustibles are found in highly inflammable pine forest where ground fires of varying intensities can occur.

Table 2 shows that the critical humidity levels for lichens and mosses along with top humus particles embedded in them does not exceed 35% in all the forest types examined. For bushes (heather, ledum, cranberry and others) this criterion is much higher—up to 60%. The explanation for this can be found in the high concentration of resins and volatile essential oils in the tissues of these plants and their position in relation to other soil cover components. If mosses and lichens were to be removed from under bushes along with the top humus, the flame would not spread along the bushes (Kurbatskii, 1962). Thus, bushes can only increase the intensity of combustion by conveying upward the flame from soil cover while the humidity of combustibles like mosses and lichens determines the inflammability of the whole complex. As a result of differential humidity of the combustibles, all soil cover components with humidity below the critical limit get burnt in forest fires.

According to the observations of S.M. Vonskii (1957), the quantity of materials burning up per unit area also depends on their humidity (Table 3).

The forest floor is an important component of the first group of

TABLE 2. HUMIDITY OF SOIL COVER COMBUSTIBLES WHICH BURN UP DURING A FIRE

Type of combustible	Humidity of cover, in % of raw weight in forest types		
	lichen-moss	moss-covered	ledum-covered
		Bushy story	
Heather	42.7–55.7	—	43.0–48.8
Ledum			52.3–59.7
Cranberry	45.4–57.0	48.4–55.5	50.0–54.0
Crowberry			43.0–56.0
		Lichen-moss story	
Lichen	7.7–34.6	—	—
Green mosses	9.4–33.0	9.3–32.3	10.0–29.0
Sphagnum			17.0–32.2

combustibles determining the intensity of combustion. The forest floor is the name given to the upper soil horizon A_0 consisting of plantlike litter in varying degrees of decomposition and with loss of natural structure. This is unlike forest top soil cover where symptoms of decomposition and destruction of structure are not observed. The decomposing forest floor forms humus.

The nature of mixture in tree species has a substantial effect on the condition of the forest floor. In soft and hard wood forest stands, the floor is not so compact as compared to pure soft wood stands; for example, those of spruce or spruce-fir.

The composition and quantity of the forest floor depend on the quantity of top humus entering the soil and the rate of its decomposition. This in its turn depends on the climate, the topography of the area, the configuration of the forest stands, their age, composition, the density of the

TABLE 3. AVERAGE WEIGHT OF COMBUSTIBLES BURNING PER 1 M^2 DEPENDING ON HUMIDITY IN PIPE STANDS OF VARIOUS FOREST TYPES

Forest type	Humidity of lichen-moss story, %	Average humidity of burnt layer, %	Average weight of burnt up materials, g/m^2
Lichen-moss	Less than 30	20	1100
	From 30 to 50	25	900
	Over 50	30	700
Moss-covered	Less than 30	20	650
	From 30 to 50	25	400
	Over 50	30	400
Ledum-covered	Less than 30	20	1300
	From 30 to 50	25	—
	Over 50	30	—

undergrowth and live soil cover. A.A. Molchanov (1960) has established that the quantity of forest floor per unit area depends on conditions of habitat and humidity. Thus, in lichen-moss pine forests, it did not exceed 2.06 kg/m^2, while in sphagnum-covered pine forests, it reached 27.68 kg/m^2 Besides, he has observed that as one crosses over northward, the quantity of the forest floor in uniform types of forests is reduced.

An important index characterizing the degree of hygroscopicity of the forest floor is its bulk weight. According to the data of A.A. Rode (1955), varies in a very wide range, from 40 to 300 g/dm^3, depending on the type of forest and exposure of the slope. N.P. Kurbatskii (1970) has determined the bulk weight of the forest floor in the main types of forest of the Karelian Isthmus: for lichen-covered pine forests—from 36 to 106 g/dm^3; for cranberry-covered pine forests—from 32 to 101; for moss-covered pine forests—from 30 to 73; for whortleberry covered forests—from 36 to 57 and for haircap-moss forests—from 31 to 47 g/dm^3. The thickness of the forest floor layer, in all types, did not exceed 5 cm. Similar results were obtained by him in calculations based on the data of A.A. Molchanov (1960). Analyzing the results obtained, N.P. Kurbatskii came to the conclusion that the more humid the forest type, the less volume weight of the forest floor forming in it, and the greater the amount of moisture it can retain. This conclusion is confirmed by the experiments of A.A. Molchanov. After 40 hours of heavy wetting, the floor humidity (calculated per unit weight in an absolute dry state) reached the following percentages: in lichen-moss pine forest—169%, in moss-covered pine forest—254%, in a whortleberry-covered pine forest—228% and in a haircap-moss forest—417%. Measurements of floor humidity carried out directly by N.P. Kurbatskii in pine forests after rainfall gave results similar to those obtained by A.A. Molchanov: in a lichen-cranberry pine forest—278%, in a steppe cranberry-herbage pine forest—198% and in a cranberry-bracken pine forest—193%. The maximum humidity after rainfall was found to be the highest in spruce forest—358%.

A characteristic feature of the forest floor is its capacity to slowly increase in humidity during the rainy period. However, once wet, it loses accumulated moisture even more slowly. The explanation for this may be found in its layered structure, its great hygroscopicity and the shading effect of the live cover. As regards the accumulation and losing of moisture, as observed by I.S. Melekhov (1947), decayed litter behaves in a manner similar to that of the forest floor. The drying process of the floor proceeds layerwise from the top and is completed only during a period of severe drought. The minimum humidity registered at the end of a drought period in the pine-whortleberry forest type was only 6%. With this humidity, the floor was burnt out completely. Under usual weather conditions, the forest floor humidity, according to the observations of V.G. Nesterov (1949), varies from 34.2% to

152.2%. At 70% humidity and above, the floor does not burn. As already mentioned, the forest floor burns mainly by flameless combustion and does not substantially affect the flame height during a fire. The calorific value of the forest floor is quite high and, according to the data of S.M. Vonskii (1957), constitutes from 4780 kcal/kg in a moss-covered pine forest to 4970 kcal/kg in a lichen-moss pine forest.

Bushes and undergrowth should be classified as surface combustibles. The most common representatives of bushes of softwood species affecting the intensity and spread of forest fires are carpeting bushes of stone pine and fir (the fir 'paw'). Dwarf stone pine in the form of undergrowth is found in many types of forests in Central and Eastern Siberia, and also in the Kamchatka and Sakhalin islands. Besides, in these regions it often forms whole thickets on watersheds and slopes, which become the main, if not the sole, combustible material. Dwarf fir as undergrowth (fir of vegetative origin) is very common in the mountains, and mainly dark-softwood forests, of southern Siberia.

The role of the undergrowth of the softwood species in relation to forest fires has not been explained so far. On the one hand, by overshadowing the soil, the undergrowth helps in moisture conservation in the soil cover, and thus seems to indirectly reduce fire risk in the forest stand. On the other hand, when fire breaks out, the undergrowth of carpeting stone pine and fir bushes often itself becomes the main combustible, which, due to its easy inflammability, sharply increases the intensity of the forest fire. Thickets of dwarf stone pine outside forest stands are always fire-prone.

The role of prominent undergrowth of the hardwood species (mountain ash, alder, bird cherry, Siberian pea shrub/caragana, spirea, rhododendron, rose bush, etc.) in the wood pyrological aspect is always beneficial, since due to a substantial green leaf mass, the undergrowth delays the spread of fire over forest areas. Besides, in spring it slows down the thawing of snow, helping to moisten the forest soil.

Everything which constitutes the upper canopy of forest stands along with tree trunks, as mentioned earlier, belongs to the second, viz. the above-ground, group of combustibles. Objects of combustion in this group are often coniferous needles and small twigs. Branches, deadwood and part of the trunks burn only in very intense fires acquiring a persistent form.

In order to study the intensity of transpiration and the photosynthesis of individual trees and the whole forest stand, the dynamics of needle accretion depending on taxonomic and other characteristics of forest stands and the determination of coniferous needle composition in the canopy of softwood forest stands was carried out by many investigators: A.V. Savina (1941), A.I. Chelyadinov (1945), A.A. Molchanov (1949, 1960) and others. The works of V.P. Molchanov (1965) and N.P. Kurbatskii (1970) are devoted to the study of the process of combustion of coniferous needles in the forest

TABLE 4. COMPOSITION OF CONIFEROUS NEEDLES AND CONE-BEARING TWIGS IN THE CANOPY OF PINE STANDS

Plot No.	Taxonomical characteristics of the stand								Composition on average tree, in kg		Composition of combustibles in canopy kg/ha (numerator—needles, denominator—twigs) with density	
	Compartment	Age, yrs	Average height m	Average diameter, cm	Density	Number of trees per hectare	Closeness of canopy	Average span of crowns, m	coniferous needles	twigs	given	reduced
						*Lichen-moss pine forests**						
1	10C	135	19	26	0.5	292	0.6	10.3	0.80 ± 0.95	1.90 ± 0.34	$\frac{2.4}{0.73}$	$\frac{3.7}{1.0}$
2	10C	125	16	24	0.5	338	0.6	7.0	7.00 ± 1.26	2.10 ± 0.60	$\frac{2.3}{0.7}$	$\frac{3.2}{1.0}$
3	10C	80	16	18	0.6	760	0.6	7.1	6.50 ± 0.80	1.50 ± 0.34	$\frac{4.2}{1.0}$	$\frac{4.9}{1.4}$
4	10C	80	11	8	0.8	4440	0.9	3.0	0.85 ± 0.29	0.23 ± 0.05	$\frac{3.9}{1.0}$	$\frac{3.5}{0.8}$
5	10C	60	10	8	0.9	5130	0.9	3.0	0.38 ± 0.04	0.11 ± 0.02	$\frac{3.3}{0.9}$	$\frac{2.6}{0.7}$
6	10C	45	10.5	10	0.6	2120	0.6	3.5	1.90 ± 0.26	0.50 ± 0.07	$\frac{2.8}{0.7}$	$\frac{3.3}{0.8}$
7	10C	40	9	8	0.8	4100	0.8	3.5	1.10 ± 0.11	0.30 ± 0.41	$\frac{3.0}{1.4}$	$\frac{2.6}{1.3}$
8	10C	35	5	5	0.8	8100	0.8	3.0	0.50 ± 0.04	0.18 ± 0.01	$\frac{4.0}{115}$	$\frac{3.5}{113}$

					*Moss-covered pine forests (cranberry)***							
9	10C	110	27	30	0.7	350	0.8	8.3	9.8 ± 2.20	3.4 ± 0.08	3.7/1.3	4.2/1.5
10	10C	80	20	20	0.6	670	0.7	6.5	7.7 ± 0.53	1.9 ± 0.07	4.0/1.0	5.3/1.4
11	10C + E	60	18	18	0.9	1200	1.0	5.8	2.6 ± 0.41	0.6 ± 0.10	4.7/1.0	4.2/0.9
12	10C + E	45	12	11	0.7	2050	0.9	4.9	1.8 ± 0.23	0.4 ± 0.07	4.0/0.9	4.6/1.0
13	10C	40	13	10	1.0	4000	1.0	3.5	1.7 ± 0.03	0.5 ± 0.03	6.8/1.8	5.5/1.4
14	10C	35	12	9	0.9	4170	1.0	3.5	1.25 ± 0.44	0.27 ± 0.10	5.5/1.2	4.9/1.0
15	10C	30	11	8.5	1.0	4270	1.0	4.5	1.20 ± 0.10	0.4 ± 0.02	6.7/1.4	5.4/1.1

*Given density 0.7
**Reduced density 0.8

TABLE 5. DYNAMICS OF CHANGE IN THE WOOD MOISTURE IN MAIN TIMBER SPECIES IN COURSE OF A YEAR

Species	Part of trunk	Moisture content, in % of absolute dry matter											
		Jan	Feb	Mar	Apr	May	Jun	Jul	Aug	Sep	Oct	Nov	Dec
Pine	Sapwood	122	116	113	115	102	110	109	100	96	115	123	123
	Heartwood	33	33	35	33	33	32	31	31	33	34	32	34
	Average	83	86	89	92	85	84	84	80	84	92	94	97
Spruce	Sapwood	145	147	142	115	106	112	110	103	112	114	117	144
	Mature wood	42	45	43	38	35	42	38	39	43	36	36	42
	Average	103	105	101	89	83	85	83	78	90	84	82	103
Birch	Peripheral	80	77	75	72	91	66	51	53	60	74	81	73
	Central	86	91	91	84	95	—	60	61	69	85	91	90
	Middle	82	86	82	76	92	70	59	60	71	78	82	84
Aspen	Peripheral	123	114	111	105	98	70	64	59	66	95	92	108
	Central	110	94	95	92	88	79	90	83	92	90	90	91
	Middle	118	107	104	102	91	72	72	64	73	91	91	105

stand canopy and to the determination of their calorific value. On the basis of investigations carried out in various regions, A.A. Molchanov (1949) has established that needle composition in fully grown forest stands varies according to age. According to him, the largest needle contents are observed during the polewood stage. Needle contents decline in proportion to increase in age. He observes that there is a close relationship between needle composition and the current rate of growth in a forest stand. V.P. Molchanov (1965) has also mentioned this relationship. He carried out a thorough estimate of needles and cone-bearing twigs in the two most inflammable types of forest, viz. moss-covered and lichen-moss forests of different ages, but of about the same stand density (0.7–0.8). The results obtained are of special interest to forest pyrologists and, therefore, are given in detail (Table 4).

Analysis of the data, given in Table 4, shows that with the indicated densities, the highest total weight of coniferous needles and coniferous twigs is observed at the age of 80 years in the lichen-moss pine forests (6.3 t/ha) and at the age of 40–80 years in the cranberry pine forest covers (6.9–6.7 t/ha). Of the two constituents mentioned above, needle reserves are subject to the highest variation with age while coniferous twigs are subject to the lowest variation.

An important factor, which determines the possibility of ignition of the forest stand canopy during intense combustion of the soil cover, is the distance between the individual trees. This relationship has been graphically expressed (Fig. 5) by V.P. Molchanov (1957). With growing age, the distance between the soil and the lower limit of the canopy in both forest types increases. The rate of increase in moss-covered pine forests, however, is higher than that in lichen-moss pine forests.

Comparison of the data in Table 4 and Fig. 5 leads to the conclusion that

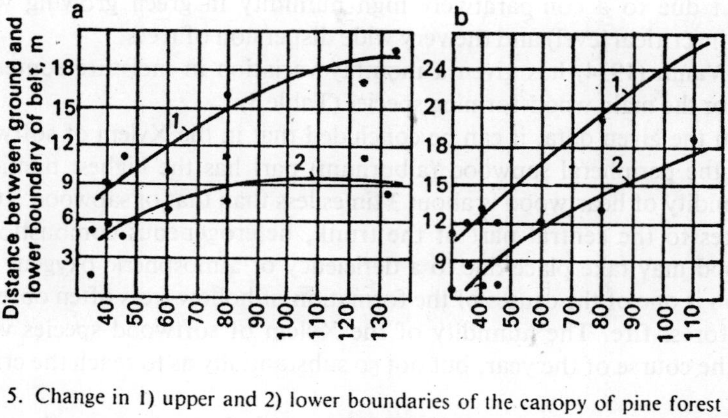

Fig. 5. Change in 1) upper and 2) lower boundaries of the canopy of pine forest stands with age:

a—lichen-moss pine stand; *b*—moss-covered pine stand.

with increasing age, the span of crowns (and consequently, the thickness of the combustible layer) steadily increases, but the quantity of combustibles per unit volume of the canopy (m^3) decreases noticeably. This law confirms the fact that with increase in age, the possibility of fire spreading along the canopy of pine forest stands decreases, since the density of the combustible layer decreases. Proceeding from these premises, stands of young coniferous trees should be considered the most fire-prone.

Coniferous needle content in spruce stands is 2–3 times more than in pine forests. Thus, according to the data of V.V. Smirnov (1964), in a 75-year-old spruce-sorrel forest of a density of 0.7, fresh needle contents were 29.5 t/ha, while in an absolutely dry state, they were 12.8 t/ha. According to the data of L.K. Pozdnyakov (1969), the quantity of needles in the western Sayan mountains varies: in stone-pine forests—from 2.2 to 10.6 t/ha, in fir groves—from 5.7 to 9.6 t/ha and in larch forests—from 1.8 to 4.8 t/ha. The quantity of leaves in a birch grove is 3 t/ha on the average.

The humidity of coniferous needles varies depending on the age of the forest stand, the season, the conditions of habitat and other factors. In spring, young coniferous needles have the highest humidity (above 300%), while old needles have the lowest. With age, the humidity of young needles gradually decreases, and toward autumn it is less than 150%.

Timber resources, potentially combustible, vary from 80 t/ha in a lichen-moss pine forest to 290 t/ha in a moss-covered pine forest. It should, however, be noted that even during fires of severe intensity, the trunk rarely burns out completely except in the case of dead trees. Flame destroys coniferous needles and twigs up to 4 cm in thickness and severely scorches the branches. Fire causes utmost damage to the trunk in places of resin outflow (in tapping) and old fire-affected and partially dried spots. Tree trunks do not burn due to a comparatively high humidity in green growing wood (above the critical level) and the very wide dispersion of trees.

S.I. Vanin (1934) has given a monthly variation in moisture content in timber for the main wood-forming species (Table 5).

From the given data, it can be concluded that in the Xylem of softwood species, the peripheral sapwood (alburnum) part has the highest humidity. The humidity of heartwood is about 3 times less than that of sapwood. If fire penetrates to the central part of the trunk, heterogeneous combustion of heartwood may take place due to a deficiency of atmospheric oxygen. This condition is one of the causes of the formation of hollow trees often observed after a forest fire. The humidity of the Xylem of softwood species varies during the course of the year, but not so substantially as to reach the critical level.

In some softwood species (pine, larch) which form the most fire-prone types of forest, protection against fire is afforded by the bark, or rather by the outer layer called the crust. The crust is a bad conductor of heat, and, due to a high content of ash elements, either does not fully ignite or burns very slowly. The crust rises up to about ⅓ of the height of the tree and gradually tapers off. Variation in the thickness of the crust in proportion to its rise along the trunk well reflects the temperature distribution according to height in forest fires.

On analyzing the growth rate of pine and larch forest stands, we may observe ecological and morphological features which enable these species to survive and increase their area in spite of periodically recurring fires.

Young saplings of pine and larch have drooping crowns and comparatively thin bark without a defensive crust layer. However, during the sapling stage, these species, as a rule, form very dense forest stands. Under these conditions, lichens and grasses, which constitute the most fire-prone components, do not develop well under the shade of saplings, and the soil cover consists mainly of top humus, namely a thin layer of dead needles of medium humidity. Under such conditions, the risk of fire spreading from below to the crowns of young trees is considerably reduced.

Pine and larch stands thin out with age and, as has been noted earlier, combustible reserves in soil cover also increase. These, in turn, increase the intensity of combustion and height of the flame. However, with age, the distance from the ground to the lower crown limit also increases, which renders the crown inaccessible to ground-fire flames while the lower part of the trunk is covered by a protective crust. In the process of natural selective growth over the ages, defensive properties of pine and larch have developed under the effect of forest fires. Therefore, it may be said that the present structure of pine and larch trees as seen growing today, has been mainly formed by fire.

Protective adaptability against forest fires can be seen also in spruce. Its drooping crown overshadows the soil cover, increasing its humidity. Besides, spruce needles, due to their much higher ash content as compared to pine, burn more slowly. Spruce saplings with their density and thickness can, under certain conditions, well withstand forest fires.

In soils with excess moisture and bogged up areas peat horizon* which represents a basis for the third group of forest combustibles, develops well. The most common type of forest growing on peat soils is the ledum-covered pine forest of IV–V classes. However, in drained areas, forest stands with higher productivity can also grow. Thickness of peat in ledum-covered forest varies from 15 to 60 cm. Usually, the whole peat mass is reinforced with a large quantity of roots of trees and bushes. The volume weight of peat soils is about the same as that of the forest floor of moss-covered stands. According to the data of M.P. Elpat'evskii (1970), it varies from 40 to 260 g/dm^3 and depends mainly on the degree of decomposition.

The calorific value of peat is quite high and, according to the data of S.M. Vonskii (1957) obtained for ledum-covered pine forests, reaches 5000 kcal/kg.

Humidity of peat during a fire-risk period falls gradually from a maximum in spring to a minimum in autumn. This characteristic is confirmed by the established law that fires occur in peat soils at the end of summer (August–September) and never during spring. Being isolated from the atmospheric layer of the forest floor, peat reacts poorly to weather changes, and only prolonged drought or rainy periods can have a marked effect on its humidity.

CALORIFIC VALUE OF FOREST COMBUSTIBLES

During the process of combustion, a liberation of heat occurs simultaneously with formation of combustion products. The quantity of heat, which is liberated on complete combustion of one kilogram of fuel in an absolutely dry state, is called the gross calorific value. In the combustion of moist fuel (for example, wood), a part of the heat is used in the evaporation of water present in it. Therefore, in this case, the quantity of heat liberated during combustion will be less than in combustion dry wood. This is called the net calorific value of the fuel. Thus, gross calorific value is higher than net calorific value by the quantity of heat used in evaporation of moisture. The unit of calorific value is a calorie, being equal to the quantity of heat required for heating one kilogram of water by 1 °C in the temperature range of 14.5 to 15.5°C.

The gross calorific value Q_h of any combustible can be determined by the calorimetric method or calculated empirically on the basis of its elementary composition. For wood, it can be calculated according to the formula

$$Q_h = \frac{8137\ C + 3418\ O\ (H - O)}{100},$$

where C, H and O represent carbon, hydrogen and oxygen present in dry wood, in percentages.

For determining the calorific balance in forest combustion, the net or utilizable calorific value Q_l is more important, viz. the one which corresponds to the humidity directly present in the combustibles of the forest. The rate of preparation of combustibles for ignition and, consequently, the rate of fire spread along the forest floor depends on the quantity of heat liberated. For wood, it can be calculated from the formula

$$Q_l = 81\ C + 246\ H - 26\ O - 6\ W,$$

where W is the percentage of moisture in wood.

In practice, the calorific value of various combustibles is determined not by formulas mentioned here, but by means of a special apparatus called a

*Hereinafter peat layers will be examined along with forest cover, considering them as a soil horizon since the compositions and degrees of decomposition of peat in peat beds of boggy soils are different.

calorimetric bomb. On igniting the sample in the bomb, the liberated heat is transferred to the water of the calorimetric flask, the temperature of which is measured with great accuracy. A part of the heat is also spent on heating the whole calorimetric system. Both the heat measurements being known, it is easy to determine the total.

A number of works (Vonskii, 1957; Amosov, 1958; Gorbatov, 1963; Telitsyn and Sosnovshchenko, 1969 and others) have been devoted to the study of the calorific value of forest combustibles. The most complete data on the gross and net calorific values of the main types of forest combustibles were obtained by G.A. Amosov (1957).

As seen from the data in Table 6, no great difference in calorific value is observed in various combustibles. Ledum leaves have the maximum calorific value (5800 kcal/kg) apparently due to a high content of resins and essential oils (about 17%). The minimum calorific values is observed in ferns. The transfer of fire to a ledum-covered forest stand results in a sharp increase in the intensity of combustion, accompanied by liberation of dark thick smoke, containing combustion products of essential oils. If the spreading fire encounters forest stands where the cover is dominated by ferns, its rate of advance is noticeably reduced. The effect of resin content in the tissues of wood on its calorific value was observed by S.I. Vanin and M.I. Ezupov (1934). According to their data, the calorific value of healthy pine with normal resin content is considerably lower than that infected by the fungus *Peridernium pini,* which intensifies resin secretion as a defensive reaction (Table 7).

The data in Table 7 indicates that the gross calorific value of pine wood is proportional to its resin content which possesses a fairly high calorific value, namely 8300 kcal/kg, much higher than any forest combustible. That is why areas of excessively tapped pine forests or forests affected by the fungus *Peridernium pini* are most fire-prone, I.S. Melekhov (1948) observes that the resin content of pine in a burnt-out zone increases in 5 years from the day of the fire, from 14 to 31%, and, in 25 years, even higher, i.e. up to 37%.

The resin content in one cubic meter of wood in the softwood species differs very widely and varies in the following range: in pine it is 48 kg, in larch—42 kg, in spruce—17 kg, and in fir—10 kg. Fir has the lowest resin content in its wood. However, its needles contain the highest quantity of turpentine (19%), while pine needles contain 13.6% and Swiss stone pine contains 11%. Therefore, fir needles are most fire-prone.

Wide variations in the net calorific value of various combustibles are observed as a result of changes in their humidity. According to the data in Table 6, the minimum humidity in most components corresponds to humidity after prolonged drought. The maximum humidity for the main types of combustibles (lichens, mosses, forest floor and peat) is fixed at the

TABLE 6. CALORIFIC VALUE OF VARIOUS TYPES OF FOREST COMBUSTIBLES

Types of forest combustible materials	Calorific value, kcal/kg		
	Gross Q_h	Net Q_1 (in numerator) with humidity, % (in denominator)	
		minimum	maximum
Pine:			
young needles	5190	3160/54	1330/200
Ledum-covered pine forest:			
needles	5530	—	—
bark	4820	—	—
wood	4930	2140/102	1460/168
Spruce:			
needles	4930	2610/67	1630/150
bark	4350	—	—
wood	4800	2630/67	1890/117
twigs	4520	—	—
Birch:			
bark	5480	—	—
wood	4800	2690/64	1070/223
Juniper	5180	—	—
Lichens:			
Cl. rangifilina	4270	4180/3	2910/40
Cl. alpestry	4380	—	—
Shreber's moss	4760	4410/7	3230/—
Heather	5360	3200/57	2060/124
Cranberry	5150	2820/68	1330/198
Ledum:			
whole	5420	2940/70	1830/148
leaves	5800	—	—
stems	5200	—	—
Top humus (mainly dead pine needles and bark)	5180	—	—
Veronica	5700	—	—
Dry ferns	4110	—	—
Haircap moss	4610	—	—
Forest floor	4340	4060/6	2930/40
Peat	5030	3150/50	420/450

TABLE 7. RESIN CONTENT AND GROSS CALORIFIC VALUE OF PINE WOOD, BOTH HEALTHY AND INFECTED BY FUNGUS *Peridernium pini*

Wood	Resin content, %	Gross calorific value kcal/kg	Gross calorific value %
Healthy	2.4	4872	100
Infected	13.5	5323	109.26
	20.0	5645	115.8
	45.3	6253	127.34

level at which they are still able to initiate combustion (flameless for peat). For other components, the humidity level is fixed irrespective of the possibility of independent combustion.

An increase in the humidity of lichen and in the forest floor from 3–6 to 40% lowers the calorific value of these heterogeneous combustibles by almost 30%. However, as mentioned by G.A. Amosov (1958), a substantial reduction in the net calorific value of one kilogram of fuel with increased humidity does not result in an equally substantial fall in the calorific value of the whole mass of fuel in a definite forest area. The quantity of the dry part of the fuel with variations of humidity in the area remains practically constant, and the consumption of heat in evaporation of excess moisture is only about 6%, when the increase in humidity is 50%. A reduction in the intensity of the forest fire with increased humidity in the combustible material is mainly the result of the nonavailability of a part of the material for the combustion process due to increased humidity, and is not due to reduction in its calorific value.*

In forest fires, our concern is not with regard to the combustibility of any particular type of combustible, but with a complex of various components in their most varied combinations. Therefore, for solving practical problems, for example, determining the quantity of retardants for quenching a fire, it is essential to know the total calorific value of the whole complex of combustibles represented in a particular type of forest. For solving this problem, S.P. Vonskii (1957) studied the relationship between the total calorific value of combustible material of soil cover in an absolutely dry state and the age of the stands of three types of pine forest: lichen-moss, moss-covered and ledum-covered. As was expected, in the lichen-moss pine forest, where the weight of soil cover increases with age, the calorific value has a rectilinear dependence and varies from 3500 kcal/kg in the 2-year old growth to 6500 kcal/kg in the 7-year old growth. For other types of forests, a nonlinear dependence was established. It has not so far been possible to establish a definite relationship of total calorific value of a combustible with the age of the forest stand.

*Amosov, G.A. Nekotorye osobennosti goreniya pri lesnykh pozharakh (Some Characteristics of Combustion in Forest Fires). Leningrad, 1958, p. 24.

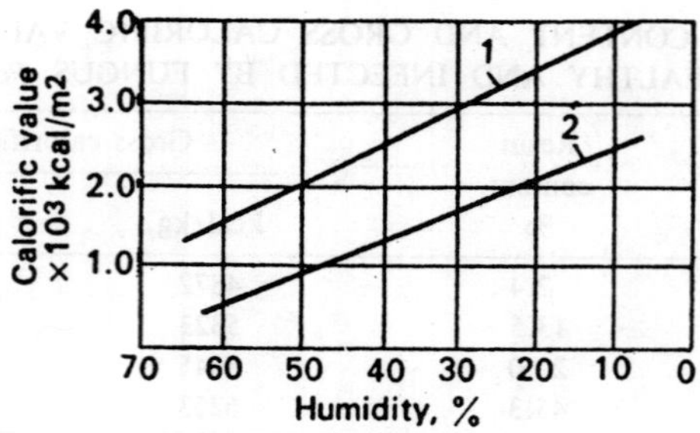

Fig. 6. Change in the calorific value of soil cover depending on its humidity:
1—lichen-moss pine forest; *2*—moss-covered pine forest.

Relationship of the total calorific value of the whole layer of combustibles with their humidity is of great interest in the application of forest control measures. This relationship is expressed by a straight line (Fig. 6), which indicates that in proportion to the fall in humidity of the lichen-moss story, the total quantity of heat liberated per square meter of area during combustion of soil cover gradually increases from 2000 to 4000 kcal/kg. In moss-covered pine forests, these values will be lower, being 1000 to 3100 kcal/kg, as a result of smaller combustible contents per unit area, and their higher humidity. This data is, of course, tentative, since the duration of the post-fire period when an accumulation of organic mass of soil cover takes place, was not considered in the experiment.

CLASSIFICATION OF FOREST FIRES

Man often encounters the phenomenon of combustion in the form of bonfires, total fire clearings in agricultural land, during removal of timber from felling areas, individual fires, etc. But not all of these can be included in the category of forest fires. A forest fire is any unexpected or unregulated combustion of vegetation spreading over a forest area.

I.S. Melekhov (1947) considers forest fires as a variety of landscape fires, which, apart from true forest fires, include steppe, tundra, meadow, swampy and other fires. He has worked out a detailed classification of all forest fires, based on the type of combustibles which burn in them. According to this classification, all forest fires are into the following types and varieties: ground fires (litter-humus, surface undergrowth and bushes, deadwood and stumps); tree-crown fires (top branches, entire tree) and underground fires (peat).

N.P. Kurbatskii (1970), when developing the basic aspects of this classification, suggested the division of fires into simple and complex, as well as single-source and multisource.

At present, the practice of forest fire control depends on the classifica-

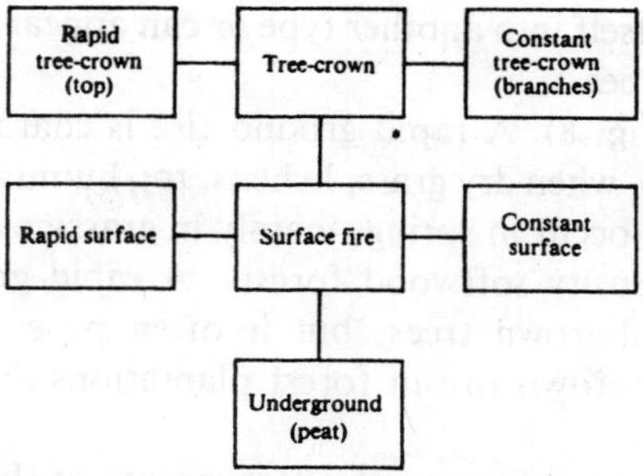

Fig. 7. Classification pattern of forest fires.

Fig. 8. Ground fire in mixed-age heather-cover pine forest.

tion pattern, the basis of which is assumed to be a particular group of combustibles burning in the fire. According to this pattern, all forest fires are classified into three types: surface, tree-crown and underground fires (Fig. 7).

In the beginning, almost every forest fire takes the form of a ground fire and, if suitable conditions develop, is transformed into an underground or tree-crown fire. Thus, according to the causative classification, ground fire can be called the main type of forest fire while underground and tree-crown fires can be called derived fires. According to the rate spread over the forest area, ground and tree-crown fires are divided into rapid and constant fires. In the process of its development, one type of forest fire can quickly transform itself into another type or can appear in a combination of two or even three types.

Ground fires (Fig. 8). A rapid ground fire is characterized by a quick advance of the edge, when dry grass, lichens, top humus, etc. burn up.

Such fires often occur in spring, mainly in grass-covered forests, and in summer in poor density softwood forests. A rapid ground fire does not usually damage full-grown trees, but it often poses a danger of being transformed into a crown fire in forest plantations and among softwood saplings.

In a constant ground fire, combustion occurs of the soil cover, stumps and deadwood. Undergrowth and bushes perish; the lower parts of trunks and roots sticking out of the ground are often damaged. After prolonged drought, the forest floor often catches fire. In a constant ground fire, the edge advances slowly emitting thick smoke indicative of the heterogeneous character of combustion. Such fires are typical of the second half of summer and are encountered mainly in whortleberry-covered and haircap-moss forests predominated by green-moss cover.

Tree-crown fires (Fig. 9). Tree-crown fires include those in which the upper-story tree crowns burn. As already mentioned, this type of fire should be considered as being derived from a ground fire, since no cases of combustion of tree canopies without the support of a ground fire have been established.

Two types of crown fires have been distinguished, namely rapid and constant. In rapid (topwood) fire, the flame spreads unevenly along the forest stand canopy, but in a staggered manner, extending along the direction of the wind. In 8–12 seconds, the flame covers a distance of nearly 100–120 m along the canopy, when its movement is suddenly reduced, and, for a few minutes, the upper story (canopy) burns, while after some time, the lower story (soil cover) also catches fire. Combustion of vegetation of the lower story is intensified due to burning twigs and needles falling from the upper story. This is why after some time the edge of the ground fire outstrips that of the crown fire and merges with individual areas of combustion which, as a rule, occur some distance ahead of the fire due to sparks. The ground fire, intensified by wind, gathers strength and heats adjacent parts of the tree canopy. With gusts of wind, the flames of the crown fire gain spasmodically along the canopy with great speed somewhat outstripping the ground fire. This is a rough pattern of development of a rapid crown fire.

Rapid crown fires are characteristic of the first, as well as the second half of summer and occur in pure pine forest stands of average density with patches of coniferous undergrowth, predominated by heather or ledum cover. In spruce and pine stands with a substantial admixture of birch, such fires are rarely observed. This condition is of great importance in laying fire-resistant forest stands.

Fig. 9. The moment of transfer of severe ground fire into crown fire.

In constant crown (entire) fires, the combustibles of the upper and lower stories of the forest stands burn simultaneously, and there is no spontaneous advance of the flame along the canopy without a supporting ground fire. In a constant crown fire, the flame advances comparatively slowly like a solid wall. This is the most destructive kind of fire. The flame of a constant fire consumes not only needles and twigs, but also branches, tree-tops, bushes and saplings, while the soil cover and forest floor are often gutted, down to the mineral layer.

Constant crown, as well as constant ground fires often occur in the second half of summer in windless weather after prolonged drought. They are characteristic of middle-aged pine and larch forests. However, if in its course, a fire encounters a small portion of hardwood, its spread is not halted and, generally, the hardwood stands are consumed in the fire.

Underground fires. Characteristic of this type of fire is the fact that the organic part of the soil is destroyed as a result of combustion (Fig. 10).

Underground fires are the result of ground fires or crown fires. Following a prolonged drought ground fires can engulf forest areas with peat soils. After the upper soil cover is burnt, smoldering continues in patches covering dry earth mounds and small heaps at the base of deadwood stumps and tree roots, where the soil is most friable and dry. With further combustion, the fire penetrates into the peat horizon of the soil, burning out funnel-shaped pits and then spreads in a horizontal manner. In practice, these are called peat fires. Since the root-holding layer of soil burns in a peat fire, trees are deprived of root support, and are uprooted (with the tree crowns usually toppling in the burnt area).

Fig. 10. Soil fire in a mixed forest stand on peat soil.

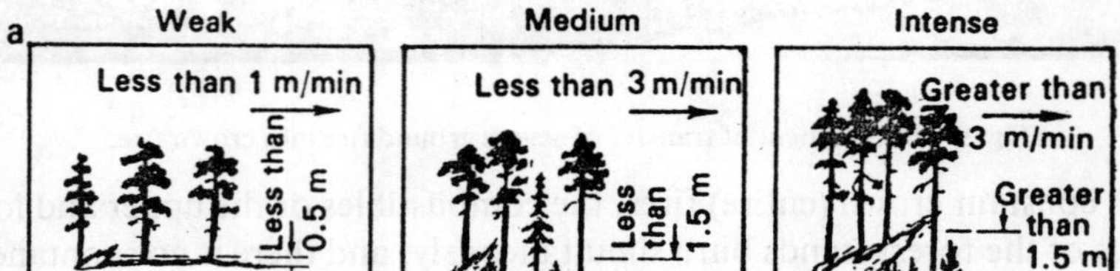

Fig. 11. Division pattern of fires according to the rate of spread and depth of burning: *a*—ground; *b*—tree-crown; *c*—soil fire.

Plans for forest fire control are worked out not only on consideration of the type of fires, but also the nature of forest areas on which they are to be implemented. Keeping this in view, all the above mentioned types of forest fires are subdivided into fires: a) under a forest canopy; b) among saplings;

c) in felling areas (cleared and with debris); d) in peat bogs; and e) in silk-worm moth habitats and forest conflagration areas.

The most important features of fires, which determine the technique of control, are the rate of spread in ground and crown fires and the burning depth in underground fires. N.P. Kurbatskii (1970) suggests division of forest fires according to these features into weak, medium and intense (Fig. 11). The rate of spread of weak ground fires does not exceed 1 m/min, that of medium fires—1 to 3 m/min and that of intense fires, over 3 m/min. A weak crown fire has a rate up to 3 m/min, medium—up to 100 m/min and an intense crown fire has the rate of over 100 m/min. According to this classification, it is recommended that we consider a peat fire to be weak, in which the burning depth does not exceed 25 cm, a medium fire has a burning depth from 25 to 50 cm and an intense fire a burning depth of over 50 cm.

The extent of combustion depends on the condition and composition of combustibles, the strength of the wind, the surface slope, the time of the day, etc. Therefore, even in one and the same fire, the rate of spread of the fire in a forest area can vary sharply. In order to correctly estimate the fire risk and to choose effective means of control, the rate of spread is established on the basis of spread in the area of highest intensity.

Effect of Various Factors on the Intensity of Combustion and the Rate of Spread of Forest Fires

The intensity of combustion in a forest is affected by many factors: nature of the combustibles (structure, volume, weight, calorific value, humidity), meteorological conditions (wind velocity, temperature and humidity of the atmosphere, solar intensity, rainfall) and even the topography (angle of inclination of the area to the horizon). The determining factor is the humidity of the combustibles, since combustion is possible only at a definite level of humidity. The other factors affect the intensity of combustion, directly or indirectly (by change in humidity), but cannot prevent it.

In heat engineering and fire-fighting technology, intensity of combustion is determined by the quantity of fuel burning in a unit of time (kg/s) or by the quantity of heat liberated during combustion per unit of time (kcal/s). S.M. Vonskii (1957), after a detailed study of the effect of several factors on the intensity of combustion in forest ground fires, considers the latter as the main index for determining the intensity of combustion at the edge. This index in turn is closely related to the height of the flame, the depth of the edge and its rate of advance. According to him, the main factors, which are related to the nature of the combustion phenomenon and most actively affect its intensity, are the calorific value of combustibles, their moisture content and wind velocity. Results of the study of the flame intensity of ground fires depending on these factors in the three most inflammable types of pine forests are given in Table 8.

According to the data in Table 8, it follows that the quantity of heat liberated per unit length of the fire edge in moss-covered pine forests under the same wind velocity and humidity of the combustibles is 3.7 times lower than in ledum-covered pine forests and almost three times lower than in lichen-moss pine forests. The ratio of rates of advances of the fire fronts under the same conditions for these types of forests will be 1 : 1.8 : 2, and of the flame height—1 : 1.5 : 2.0.

Wind is an important factor since it increases the intensity of combustion of the frontal edge. With the same humidity of combustibles in various types of forests, due to wind, the increase in the quantity of heat liberated per unit of time is 6–25 times, the rate of advance of the edge, 3–19 times, the depth of the edge 3–12 times, and the flame height 2–5

TABLE 8. FLAME INTENSITY OF THE FRONT OF GROUND FIRES IN RELATION TO MAIN FACTORS

Wind velocity, m/s	Factors of combustion with humidity of lichen-moss story, %											
	up to 30				from 30 to 50				above 50			
	$q \times 10^3$ kcal	V, m/min	h, m	l, m	$q \times 10^3$ kcal	V, m/min	h, m	l, m	$q \times 10^3$ kcal	V, m/min	h, m	l, m
					Lichen-moss pine forest							
0.5	2.6	0.7	0.5	0.5	2.0	0.6	0.3	0.4	1.0	0.5	0.2	0.2
1.5	8.5	2.1	1.4	1.9	5.2	1.7	1.0	1.2	2.3	1.1	0.6	0.7
2.5	16.0	3.9	1.8	2.9	8.8	2.9	1.4	1.9	4.3	1.7	1.1	1.1
3.5	23.5	5.8	1.7	3.6	12.4	4.0	1.3	2.5	6.6	2.3	1.1	1.5
					Moss-covered pine forest							
0.5	1.3	0.5	0.6	0.5	0.2	0.3	0.4	0.3	—	—	—	—
1.5	3.0	1.1	1.0	1.0	0.6	0.6	0.5	0.4	—	—	—	—
					Ledum-covered pine forest							
0.5	3.0	0.7	1.3	0.9	—	—	—	—	—	—	—	—
1.5	11.1	2.3	2.0	1.8	—	—	—	—	—	—	—	—
2.5	20.9	4.3	2.3	2.7	—	—	—	—	—	—	—	—
3.5	30.7	6.3	2.0	3.5	—	—	—	—	—	—	—	—

q—quantity of heat liberated per linear m of fire edge per min, in thousand kcal,
V—rate of advance of front, m/min,
h—flame height, m, and
l—depth of edge, m.

times. In ledum-covered and lichen-moss forests, where a second story is absent, the effect of wind on the intensity of combustion is more pronounced.

Under the canopy of forest stands, wind velocity depends on the structure, height and the density of the canopy (Table 9). But, the composition of the forest stand, i.e. the presence of a particular growing stock, has the greatest effect on wind velocity in a forest. An example of such an effect with the same height and stand canopy can be seen from the experimental measurements carried out by E'.N. Valendik (1968) in two types of forests, viz. moss-covered spruce and cranberry-covered pine.

TABLE 9. EFFECT OF STAND COMPOSITION ON THE BEHAVIOR OF WIND VELOCITY

Types of forest	Compo-sition	Height, m	Canopy den-sity	Wind velocity, m/s with observation height above ground surface, m			
				21.4	13.6	10.2	5.1
Moss-covered spruce forest	9EIB	17	0.7	5.5	1.1	0.3	0.6
Cranberry-covered pine forest	10C	17	0.7	5.2	0.8	0.6	0.8

In both types of forests, the wind velocity decreases in the canopy by 5–6 times and then slightly increases under the canopy. The general description of wind behavior depending on the forest stand composition is given in Fig. 12. This data confirms the wind profile first worked out by R. Geiger (1931) according to measurements taken at various heights from the soil surface in a pine forest. With wind velocity of 1.61 m/s over the crowns

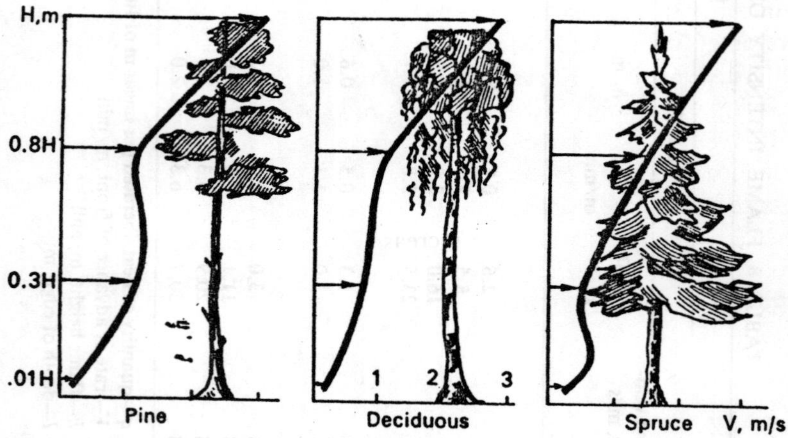

Fig. 12. Changes in wind profile in relation to the composition of forest stand.

of a mature pine stand of 17 m height, velocity at the height of 1.1 m above the soil surface was almost 3 times less (0.6 m/s).

V.P. Molchanov (1961) repeated the observation of R. Geiger in a similar pine forest stand. According to his data, when the wind velocity was 6 m/s above the crowns, it was about 2 m above the forest cover. He also arrived at the conclusion that the structure of the forest stand is the main factor governing wind behavior in forests. A.A. Molchanov (1961), on conducting wind velocity measurements in an open space in a forest, arrived at the conclusion that the wind velocity in a forest is 40–60% of that in an open space. Just above the soil surface, it is even lower among lichen-moss cover and bushes.

It is very important to determine the wind velocity in forests at various heights in relation to velocity measurements at the meteorological station for the application of forest fire control. According to the data of S.M. Vonskii (1957), wind velocity in pine forests is, on an average, 6 times less than that at the meteorological station. For pine forests of various heights, E'.N. Valendik (1968) found a mathematical relationship between these values which can be expressed by the linear equation $y = a + bx$, where y is the wind velocity in the pine stand at a certain height, in % of x; and x the wind velocity at the meteorological station, at a distance of not more than 15 km from the stand.

Wind also intensifies evaporation and thereby helps in lowering the humidity of forest combustibles. But the role of wind in the spread of fires is especially significant. The direction and velocity of spread of forest fires is mainly determined by wind. By directing the flame and hot air current toward the soil cover, wind intensifies the heating of combustibles and, as a result, hastens their spontaneous ignition. Wind causes spread of rapid ground fires and aids in converting intense ground fires into crown fires.

The effect of wind velocity on the rate of advance of the fire edge as well as increase in its area and perimeter was studied by many investigators. Several of them (V.G. Monakhov, 1964; I.V. Ovsynnikov, 1965; G.P. Telitsyn, 1965; N.P. Kurbatskii, 1966) composed formulas for determining areas and perimeters of fires according to known linear fire measurements.

In 1968, G.N. Korovin investigated the propagation of linear velocities of tactical elements of forest ground fires in heather- lichen-moss- and moss-covered pine forests and established a mathematical relationship between wind velocity and the increase in linear velocity of the fire front. This relationship (under constant moisture content of combustibles) is expressed by the following equation:

$$(x) = a_0 + a_2 x^2,$$

where a_0 is a constant, numerically equal to the velocity of propagation of fire in the absence of wind; a_2 the coefficient of proportionality between the

TABLE 10. VALUE OF COEFFICIENT OF PROPORTIONALITY DEPENDING ON THE TYPE OF FOREST AND HUMIDITY OF COMBUSTIBLES

Types of forest	Proportionality coefficient with humidity of combustibles, %	
	10–20	20–30
Pine forest:		
heather-covered	0.760	0.546
lichen-moss	0.675	0.389
moss-covered	0.525	0.342

square of wind velocity and the mean spread velocity of the fire front; and x the wind velocity in m/s.

Numerical values of the coefficient of proportionality, a_2, vary for each type of forest and depend on humidity of the combustibles (Table 10).

With increased humidity of combustibles, the numerical values of the coefficients a_0 and a_2 are reduced.

The degree of correlation between theoretical values of the spread velocity of the fire front and experimental data remains within the range of 0.6–0.8 and is determined mainly by the heterogeneity of combustibles within each type of forest and the nonuniformity of their distribution.

Using the principle of simulation, the essence of which lies in the fact that the fire is examined under special model conditions (which do not change throughout the period of fire action) G.P. Telitsyn (1965) derived a formula on the relationship of the rate of propagation of the main tactical fire elements (front, flank and rear) on wind velocity and the nature of combustibles, which is given in the general form:

$$v=(v_0+Kv_w)\left(1+\frac{v_w}{\sqrt{v_w^2+C^2}}\right)^2,$$

where v_0 is the spread velocity of fire in a plain in windless weather, m/min, v_w the wind velocity, m/min, K the coefficient accounting for the blow effect of the flame, and C the specific heat of combustibles, kcal/kg deg.

For practical calculations, the author suggests that the value of v_0 should be taken in the range of 0.4–0.6 m/min with humidity of the fuel up to 30%, and 0.2–0.4 m/min with humidity of over 30%. Coefficients K and C are found by experiment. Based on the data of S.M. Vonskii, G.P. Telitsyn (1965) established numerical values of these coefficients in relation to the character of combustibles (Table 11).

An example of calculation of the actual rate of advancement of the edge at the front, flank and rear of the fire is given below, with soil cover represented by dry grass and lichens with a humidity of less than 30% and a wind velocity of 4 m/s.

Following the above directions, it is assumed that $v_0=0.5$ m/min.

TABLE 11. VALUES OF COEFFICIENTS K AND C IN RELATION TO THE CHARACTER OF COMBUSTIBLES AND THEIR HUMIDITY

Type of combustibles (with average cluttering)	Coefficients	Humidity of combustibles, %		
		<30	30–50	>50
Dry grass, lichens, top humus of needles and leaves	$\frac{K}{C}$	$\frac{0.45}{3.50}$	$\frac{0.27}{3.30}$	$\frac{0.16}{3.00}$
Green mosses	$\frac{K}{C}$	$\frac{0.20}{2.40}$	$\frac{0.10}{2.20}$	$\frac{0.05}{1.80}$

According to Table 11, the values of coefficients K and C, corresponding to a humidity of up to 30% ($K=0.45$ and $C=3.5$) are determined. Substituting numerical values in the formula, the rate of advance of the frontal edge is found:

$$v_{fr}=(v_0+Kv_w)\left(1+\frac{v_w}{\sqrt{v_w^2+C^2}}\right)^2$$

$$=(0.5+0.45\times 4)\left(1+\frac{4}{\sqrt{4^2+3.5^2}}\right)^2=7\text{ m/min};$$

of the rear edge

$$v_r=(v_0+Kv_w)\left(1-\frac{v_w}{\sqrt{v_w^2+C^2}}\right)^2$$

$$=(0.5=0.45\times 4)\left(1-\frac{4}{\sqrt{4^2+3.5^2}}\right)^2=0.15\text{ m/min};$$

of the flank edge

$$v_f=v_0=Kv_w=0.5+0.45\times 4=2.3\text{ m/min}.$$

Analysis of the above formula of G.P. Telitsyn shows that there are functional relationships between the rate of advance of the front, flank and rear of fires which can be expressed by the equations:

$$\frac{v_{fr}}{v_f}=\left(1+\frac{v_w}{\sqrt{v_w^2+C^2}}\right)^2;\quad \frac{v_{fr}}{v_r}=\frac{\left(\sqrt{v_w^2+C^2}+v_w\right)^2}{\left(\sqrt{v_w^2+C^2}-v_w\right)^2}$$

These equations help in determining the velocity of propagation of any of the fire edges if the velocity of one of them is known, and consequently in correctly estimating the risk of starting a fire in any area.

As a rule, the process of combustion in crown fires, especially the constant ones, is hidden from the observer by a thick cloud of smoke and, therefore, has barely been studied so far. It is risky to conduct field experiments on this particular aspect while its simulation under laboratory conditions involves certain difficulties.

The possibility of occurrence and spread of crown fires depends on three essential factors: 1) the total calorific value of the soil cover, 2) the distance between soil cover and the lower boundary of the canopy and 3) the pyrological character of the tree canopy (needles' weight, small twigs per unit of canopy volume, chemical composition and resin content of needles and their humidity). When these characteristics are known, it is possible to make a prior determination of the probability of the extension of a ground fire into a crown fire for each type of forest stand from a knowledge of the heat budget. An example of such calculation for lichen-covered pine forests was given by V.P. Molchanov (1957).

If the fuel content per square meter of soil cover is taken as 1 kg, its calorific value for the above type of forest is 5060 kcal. On the basis of experimentally established temperature tables, heat distribution during combustion of a 1 m wide edge proceeds in the following directions: downward for heating the soil—10% (506 kcal), forward and backward—15% each (760 kcal each) and upward—60% (3040 kcal). Heat dispersion occurs proportionate to the upward rise in the same ratio. Thus, out of the 3040 kcal of heat, only 720 kcal will reach the canopy. At the same time, about 220 kcal of heat is required for heating the needles and a cubic meter of air surrounding them, to ignition temperature (330°C). Under these conditions, ignition of the canopy is highly probable. However, this does not happen in practice, since even a gentle breeze under the canopy always deflects the fire head from the heated area and thereby prevents a rise in the temperature of the canopy to a critical level, when ignition is possible. A higher heat intensity per unit of time is necessary for this to occur. This condition can be fulfilled, for example, with the response of beds of coniferous undergrowth encountered in the path of movement of the ground fire edge. The total liberation of heat due to combustion of the soil cover and undergrowth and the rise of the flame to a greater height creates a real ignition threat for the canopy.

A large quantity of heat is liberated in crown fires. Very hot air along with combustion products is displaced by heavy cold air from the upper atmospheric layers, forming the so-called convection column. According to the observation of the authors in 1972 along the banks of the Angara River, due to a large all-embracing crown fire, the convection column reached a height of 5.6 km, ending in a thick cumulus cloud. According to visual observations, the diameter of the column base was over 800 m, gradually converging toward the base of the cloud. The current of heated gases rose along the 'column' at the rate of tens of meters per second, accompanied by rare flashes of unburnt decomposed products of wood. Such a spread of crown fires is observed under unstable atmospheric conditions, i.e. in turbulent air. With dense inversion layers established at a low height, the kinetic energy of a crown fire is sharply reduced.

In underground (peat) fires, peat burns through the entire depth up to the mineral layer or up to the layers with high humidity, where combustion is not possible. Usually, peat combustion occurs in a flameless phase as a result of 'oxygen deficiency', mainly due to oxygen entering along with air. Therefore, the rate of advance of a peat fire edge is comparatively slow, viz. from a few decimeters to tens of meters per day.

On examination of a sectional view of a peat fire edge, it can be seen that the process of burning up of the peat mass occurs in the lower part of the 'oven' much faster than in the upper part (Fig. 13). This is explained by the fact that being heavier, fresh air saturated with oxygen first enters the lower part of the combustion zone, promoting greater intensity of combustion in this part of the 'oven'. Then the hot air, deprived of oxygen and enriched with combustion products, rises upward, flowing around the upper part of the combustion zone and thereby obstructing the entry of fresh air into it. Due to oxygen deficiency, the process of combustion in the upper part of the 'oven' is slower.

Thus, peat fire (due to its peculiar airflow dynamics) has a constant tendency of self-penetration. On penetrating the lower peat layers, it can sometimes spread by tens or even hundreds of meters from the inlet, emerging at the surface in a few places.

The calorific value of peat is higher than that of wood and reaches 6600 kcal/kg with a comparatively low ash content (about 13%). Therefore, peat is a good fuel and is, moreover, self-restoring. With a 50 cm layer of peat 165000 kcal of heat can be liberated in combustion in 1 m^2, which is quite sufficient to prepare this layer for combustion with a humidity of up to

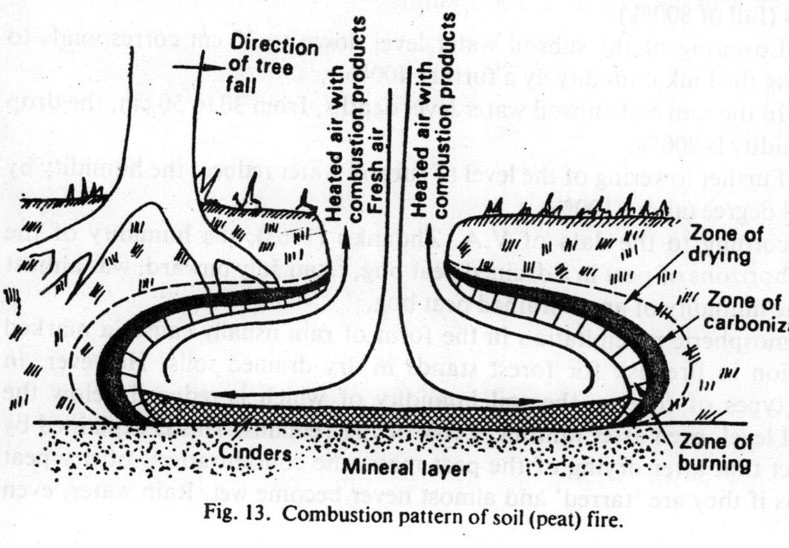

Fig. 13. Combustion pattern of soil (peat) fire.

500%, considered a critical level (Kurbatskii, 1962). The high calorific value of peat is explained by its high bitumen content (up to 25%). At high temperatures, bitumen decomposition occurs with the liberation of vaporous paraffins, which, on contact with the upper cold layers of peat, paraffinize them, i.e. envelop small particles with waterproof films. Paraffinized peat is, therefore, not moistened by water. This, long with its high calorific value, creates great difficulties in the liquidation of peat fires.

The humidity of peat is at its maximum in spring and minimum during the fire-risk period which is usually observed in August. Therefore, peat fires occur most often at the end of a summer or at the beginning of autumn. For the northwestern regions of the USSR, the probability of occurrence of peat fires is mainly determined by the quantity of rainfall during the summer. With average normal rainfall during the season (250–280 mm) in forest types with peat-covered soils, ground fires without penetration are possible. With rainfall below normal or nil in July and August, the probability of peat fire occurrence increases sharply. Such a situation, for example, developed at the end of the summer of 1972 in the central regions of the European part of the USSR, after a drought period which had lasted for nearly one and a half months.

Forest drainage also promotes the occurrence of peat fires. There exists a definite curvilinear relationship between the humidity of peat in the upper layer and the depth of occurrence of subsoil water (Konstantinov: Yuzepchuk, 1972). On studying this relationship in the draining process of bush-sphagnum-covered boggy forest massif *Otkhozhii les,* the authors arrived at the following conclusions:

1. Lowering of the subsoil water level from the soil surface to a depth of 10 cm results in a fall in the humidity of the upper peat layer from 2400 to 1600% (fall of 800%).

2. Lowering of the subsoil water level down to 30 cm corresponds to reducing the bulk humidity by a further 400%.

3. In the range of subsoil water level depths, from 30 to 50 cm, the drop in humidity is 200%.

4. Further lowering of the level of subsoil water reduces the humidity by a small degree (about 100%).

According to the data of V.A. Zhdanko (1965), the humidity of the upper horizons of peat in a drained peat bog, from July onward, was almost half the humidity of an undrained peat bog.

Atmospheric precipitation in the form of rain usually causes a marked reduction in fire risk for forest stands in dry drained soils. However, in moist types of forests, the soil humidity of which is reduced below the critical level, even after rainfall, a risk of fire remains. This is explained by the fact that after drying of the peat mass, the concentrations of dry peat look as if they are 'tarred' and almost never become wet. Rain water, even in showers, penetrates unhampered through the porous nonwettable lining of peat and indirectly reduces fire risk in the area by raising the level of subsoil water.

In certain forest stands of the taiga zone, the forest floor layer reaches a thickness of 50 cm with considerable compaction. Therefore, in constant ground fires, the combustion process is similar to that in peat fires. When thickness is insignificant (2–3 cm), the forest floor dries up quickly and burns along with the soil cover. With an increase in the thickness of the forest floor and its humidity, the combustion process can be divided into two stages. In the beginning, the ground edge passes along the surface fairly rapidly. Here, first, the soil cover (mosses, lichens, top humus) burns and then, after a considerable time lag, the edge of the constant ground fire slowly advances (secondary fire). The spread rate of the edge of such fires depends mainly on the humidity of the forest floor which can vary in the range of 250 to 6–8%. The critical humidity, at which smoldering of the forest floor is still possible, should be considered as 70%. With such humidity, the rate of advance of the edge does not exceed a few meters per day, since wind velocity has almost no effect on the intensity of combustion and movement of the fire edge. The forest floor can, however, burn with a flame in proportion to the extent of drying. Semidecomposed constituents in the form of deadwood and branches increase the combustion intensity and the role of wind in this case becomes increasingly significant.

The layer of the forest floor is usually threaded with a thick network of roots of bushes, saplings and undergrowth. Besides, the forest floor usually covers the root 'flanges' of trees which have been charred in constant ground fires. Further, the thicker the layer of the forest floor, the more severe the degree of damage to the root flanges. In constant ground fires, trees, as a rule, are not uprooted, although due to damage to root systems considerable subsequent damage to wood is possible.

Detection of Forest Fires

Detection of forest fires is mainly carried out from stationary fire observation posts, as well as by aerial and ground patrolling.

Usually, a forest fire begins from an insignificant source (a discarded burning match, a smoldering cigarette butt, a spark from a transport vehicle, a stroke of lightning, etc.) and to extinguish it in the initial stages of combustion is not very difficult. After a few hours, however, its liquidation requires many hours and hundreds of persons and even the use of heavy fire-fighting equipment.

For this reason, more emphasis is given to the early detection of fires in the functioning of a forest fire-protection service.

For better visibility in observing forests, special structures are constructed, such as fire observation towers and 25 or 35 m high masts. The basic material used in the construction of these structures used to be wood; the use of metal began only recently.

Fire observation towers are constructed in a pyramidal shape, so as to provide more stability, with the observer's cabin at the top. The following equipment is provided in the cabin: an azimuth circle with a viewing pointer for determining the location of fire by the method of intersection, a map of the protected forest area showing compartment boundaries and chief visual landmarks, binoculars, a clock, and a telephone or radio set for transmitting information about the state of fire to the fire-fighters. The azimuth is made of metal (usually aluminum) or a plastic sheet and contains 360 divisions with a clockwise numbering. The circle is fixed to a table on the top landing of the observation tower, with the zero division oriented precisely to the north.

On detecting a fire, the observer in the tower guides the pointer of the azimuth circle to the base of the smoke cloud and, according to the point of the needle, notes the azimuth angle of the direction of the fire. The observed azimuth angle is then communicated by telephone or by means of the radio transmitter to the forest division or forest enterprise office. For determining the place of fire by the method of intersection, the forest division office has a map of the forest stands with an accurate entry of all the observation towers with their azimuth circles. On receiving information about the bearings of the fire from not less than two neighboring towers, straight lines are drawn through the centers of the azimuth circles and the corresponding divisions on these circles until they intercept. The point of intersection of the lines will indicate the location of the fire on the map. If the network of fire observation towers of the delineated area is not sufficiently well distributed and the method of intersection cannot be applied, the place of fire is determined from one tower. In such a case, the map of the delineated area, also directed precisely north, is placed under a transparent azimuth circle. The center of the azimuth circle must coincide with the location point of the tower in the plan. Determination of the direction of the fire is carried out in the same manner as in the first case, but in addition to communication of the azimuth to the forest division, the observer conveys visual landmarks (tall trees, hillocks, outlines of felling areas, etc.) and their location on the map, and visually estimates the distance from the fire.

Fig. 14. Visual apparatus for determining the location of fire from an observation tower (Canada).

In Canada, an accurately leveled azimuth circle with divisions in fractions of degrees and a diopter alidade are used for more accurate determination of the location of fire from one observation tower, instead of the simple visual apparatus (Fig. 14). The structure of the apparatus permits shifting of the alidade in a direction perpendicular to the azimuth in order to avoid obstruction (for example, a cabin post). The vertical circle of the alidade helps in determining the distance of the fire from a certain point with an accuracy of up to 500 m. The range of operation of the towers is determined by the degree of visibility, and is on the average, up to 18 km.

The location of a fire can be fairly accurately determined from one tower if the surrounding terrain is photographed in the form of circular panorama and grid references marked on it. Here, the divisions of the horizontal axis should correspond with those of the azimuth circle of the alidade. After detecting the fire on the terrain, the observer traces its exact position on the photograph and then determines its position in the coordinate system. The information receiving center has an exact copy of the panorama of the terrain with the grid references. According to the coordinates communicated from the tower, it is easy to locate the fire on the photograph and, consequently, on the terrain. Such a method of fire detection is most effective in mountainous country. Moreover, the observation post here is established at the center of an elevation in order to avoid uncovered zones and ensure widest observation. In Canada and the United States, relief models of the area or stereoscopic aerial photographs are used when planning a tower network.

The foregoing broadly describes devices used for fire detection from ground observation posts, irrespective of their design features.

Fire observation towers in the USSR are constructed to cover an area of $(8–15) \times 10^3$ hectares with an effective fire detection radius of 5–7 kilometers. However, when situated on elevation dominating surrounding areas, the effective detection radius increases by 1.5–2 times.

A great advantage of wooden fire observation towers is their simplicity of structure and rigidity, as well as the availability of material for construction at site. However, they are not permanent and, therefore, unreliable. Though erected at a comparatively high cost (nearly 3300 rubles), they do not usually last longer than 5–6 years (Antsyshkin, 1957).

In order to increase the durability of fire observation towers, it is recommended that their props be created with antiseptics and they be erected on concrete foundations or on lengths of rail. Marusch (GDR), on the basis of the experiment of preservation by planking of wooden buildings, proposes planking of pyramidal tower structures with boards, thus protecting the framework from the effect of weather (S. Lange, 1963). Such protection increases durability of the wooden structures, but the expenditure on labor and wood in carrying out the project increases by 3–4 times.

One of the ways of increasing the service life of fire observation towers is the replacement of wooden by steel structures. In this case, the service life increases by 7–10 times. An example of such a structure is the 35 m high fire observation tower of the *Soyuzgiproleskhoz Institute* (Fig. 15).

The lower part of the tower (up to the mark of 6 m) has the form of a truncated pyramid 7 × 7 m in size, and the upper part is 3 × 3 m in size ending in a covered landing for the observer. The tower is built with steel angles. It is wind-resistant and is intended to withstand a dynamic wind pressure (at a height up to 10 m from ground level) of 27 kg/m^2 which corresponds to the first wind zone in the country. However, taking into account the fact that

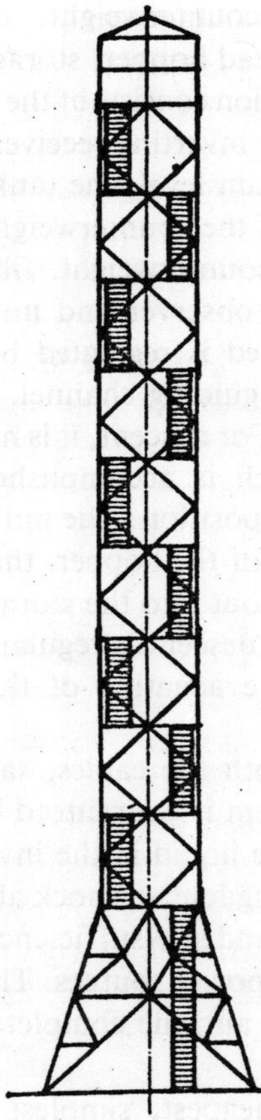

Fig. 15. Steel fire observation tower constructed by the *Soyuzgiproleskhoz Institute*.

forest stands substantially decrease wind velocity due to the uneven and porous surface of the canopy, this tower can be installed even in the second and more severe wind zone where the dynamic wind pressure in an open space reaches 35 kg/m^2. The total weight of the steel structures of the tower is 12 t, whereas the project construction cost is 3600 rubles.

To ascend the tower there is a steel staircase inside with separate sections of 3 m each, with the exception of the first, which is 6 m. The observation landing is 2.5 × 2.5 m in size and is equipped with a table and an azimuth circle. Contact of the observer with the forest division is established by means of a *Nedra* radio set or a *RSO-5 Olen'*.

The ascent of the observer to the high tower is comparatively dangerous and entails considerable muscular strength. This defect is removed in the second type of metallic fire tower, constructed by the same institute, which is equipped with unique hydraulic jack. The main components of the jack are: the observer's cabin, counterweight, block system with cables, inverting receiver tank, overhead hopper, storage tank for fluid and 2 hand pumps. The principle of operation consists of the following: before ascent, the observer pumps water into the inverting receiver tank installed at the top of the tower. On reaching a certain level, the tank inverts and the water in it flows out into the container of the counterweight. The receiver tank returns to its original position. The counterweight, filled with water, exceeds the weight of the cabin with the observer and moves downward, raising the cabin upward. The climb speed is regulated by a special brake from the cabin which moves along the guiding channel. After reaching the top, the cabin is secured by the brake. For descent, it is necessary to release the water from the counterweight which is accomplished automatically when the counterweight is in the lowest position. The pin of the valve of the counterweight is pushed into the pad of the hopper, the valve opens and the water from the counterweight flows out into the storage tank through the hopper in 2 to 3 minutes. The cabin's descent is regulated by manual brakes and is possible only after complete evacuation of the water from the counterweight.

In case of snapping of both the cables, safety during the ascent and descent with such a hoist system is guaranteed by hard brake catchers, the regulation of the weight of the liquid in the inverting receiver tank and by brakes and adjustment of spring loaded shock absorbers. A great advantage of this system is its simplicity and self-sufficiency, i.e. independence from a power-supply system or other power sources. The total construction cost of the tower, including materials and the complete equipment of the receiver tank is 4400 rubles.

Efforts to develop the cheapest, simplest and most easily operated observation posts for forestry led to the idea of using wooden single-shaft masts supporting a guy system in a vertical position instead of the usual pyramidal towers. During experiments conducted in 1955 at the Leningrad Scientific Research Institute of Forestry for testing the proposal of Forest Ranger, F.V. Bogoyavlenskii, which uses a system of four or six mirrors installed on top of the mast at an angle of 55° to the horizon gave an impetus to this idea. This method was not found to be practical since the angular dimensions of the objects reflected in the mirrors were extremely small, and the process of observation, itself, was difficult and tedious.

In 1955, the Leningrad Scientific Research Institute of Forestry conducted trial observations from a mast on which mirrors were previously fixed. These trials established that rocking of the top, which is natural in single-shaft structures, does not obstruct observation. For facility of observation, the mast should be equipped with a bridge or with a cabin for the observer on duty and a ladder or some other device for ascent of the observer to the cabin. The first such model of the fire observation mast,

Fig. 16. Fire observation mast of the *PNM-2* type, constructed by the Leningrad Scientific Research Institute of Forestry.

PNM-1, was constructed in 1958. This mast is a wooden shaft, 35 m in height, held in a vertical position by three-storied guide ropes. Along the mast is a ladder formed by driving pins into the shaft, while the latticed cabin is fixed at the top. For easy ascent of the observer to the cabin, there is a system consisting of two blocks at the top and a cable thrown across them, which is provided with a counterweight of 40 kg at one end and with a waist belt at the other. The trials of the mast once again confirmed the possibility of using single-shaft masts with guide ropes for inspecting the surrounding terrain. At the same time, it was noted that the technique of ascent to and descent from the mast does not meet the safety requirements.

In 1959, a better perfected mast structure of this type was developed, viz. the fire observation mast *PNM-2* (Fig. 16). As compared to the first model, it has a closed cabin, but the fundamental difference between it and *PNM-1* model is that the ascent device is a closed balanced two-cable elevator system, consisting of a cradle cabin with a set of adjustable ballast weights and a counterweight equal in weight to a cradle cage with a man in it. By varying the ballast weights, it is easy to achieve equilibrium in the system during the descent of an observer weighing 50 to 90 kg. The two-cable system, with speed limiting devices, ensures the required safety for ascent to any height (Speranskii, 1961).

Experience acquired in the operation of fire observation masts has shown that during the fire-risk period, continuous forest observation is not essential. Inspection of the surrounding terrain once every hour is sufficient. Consequently, the cabin may not be fixed at the top of the mast, and the man on duty can periodically inspect the surrounding area without leaving the elevator cage. Such a development was most desirable, since it helped in substantially curtailing the metal content of the mast and imparted a high safety margin to its structure by avoiding the dynamic wind forces, which occurred as a result of the 'sailing' behavior of the cabin.

The prototype of the *PNM-3* mast was produced in 1961, taking into account the above-mentioned structural changes, and was installed in the area of jurisdiction of the Northern Forest Division of Leningrad Province. The fire observation mast, *PNM-3,* is also a single-shaft structure installed on a concrete foundation and held in a vertical position by a system of guide ropes (Fig. 17). The mast is provided with a suspended self-operated elevator including a cabin. The highest point, from which observation is possible, is at a height of 35 m above the ground. The shaft of the mast is made by joining three pine logs. In order to obtain the specified diameter for the mast, the two lower logs are joined at the butt ends. The logs are held together at the joints by two metal sockets built in at the base. A metal axle, passing through the hole in the shaft at the base and held by the sockets, supports the shaft. To avoid crushing and cracking of the lower part of the mast near the butt end, by the axle, immediately after the climb, two wooden bars are interlaid, and the shaft is fastened with a strap at the base (Shtuchkov, 1965). The ladder of the mast shaft is made of metal. It consists of two stringers connected with rungs situated at a distance of 0.5 m from one another. The stringers are made of unequal angle bars and serve as guides for the self-operated elevator. The ladder is fastened to the mast shaft by means of pins welded to the inner sides of the stringer.

The guide ropes of the mast are fixed at three levels. As would be seen from an examination of the plans, the guide ropes at each level are adjusted at an angle of 90° to one another and joined to the straps of the mast shaft by loop shaped ends in which eyelets are interlaced. The guide ropes of the lower and middle level are fastened to the middle straps connecting the corresponding butt joints of the shaft, while those of the upper level are fastened to the strap situated directly under the rigging of the top. The lower ends of all the guide ropes are fastened to Roman nuts by screw clamps. The Roman nuts, in their turn, are joined by shackles to four

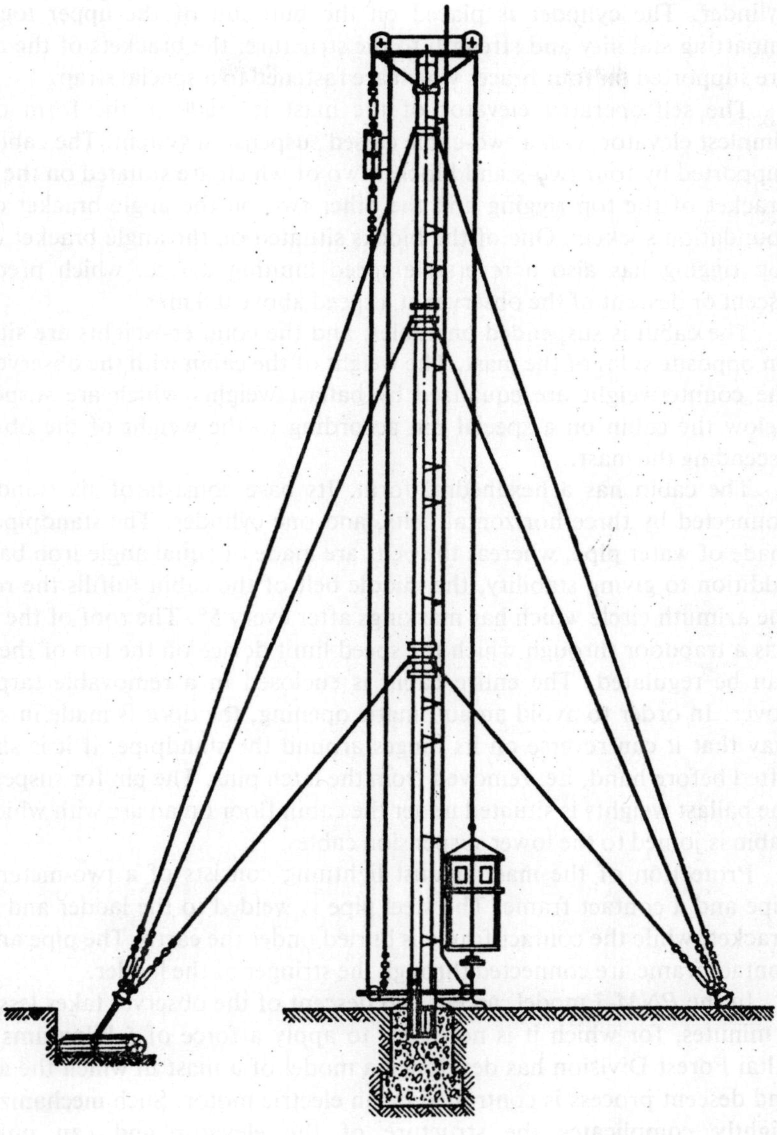

Fig. 17. Diagram of fire observation *PNM-2* mast, constructed by the Leningrad Scientific Research Institute of Forestry.

metallic heavy-duty braces. The metal braces are connected to wooden stays, which are buried 2.5 m deep—forming an anchor. Thus, the three guide ropes are joined to each anchor, one from each level.

At the top of the mast, a shaft is fastened to a double-ended angle bracket, which consists of two angles joined by two link plates and a metal cylinder. The cylinder is placed on the butt-end of the upper log. For imparting stability and strength to the structure, the brackets of the angles are supported on four braces which are fastened to a special strap.

The self-operated elevator of the mast is made in the form of the simplest elevator with a two-cable closed suspension system. The cables are supported by four two-stand blocks, two of which are situated on the angle bracket of the top rigging and the other two on the angle bracket of the foundation sockets. One of the blocks situated on the angle bracket of the top rigging has also a reversible speed limiting device, which precludes ascent or descent of the observer at a speed above 0.8 m/s.

The cabin is suspended on cables, and the counter-weights are situated on opposite sides of the mast. The weight of the cabin with the observer and the counterweight are equalized by ballast weights which are suspended below the cabin on a special pin according to the weight of the observer ascending the mast.

The cabin has a hexahedral form. Its base consists of six standpipes connected by three horizontal belts, and one cylinder. The standpipes are made of water pipe, whereas the belts are made of equal angle-iron bars. In addition to giving stability, the middle belt of the cabin fulfills the role of the azimuth circle which has markings after every 5°. The roof of the cabin has a trapdoor through which the speed limit device on the top of the mast can be regulated. The entire cabin is enclosed in a removable tarpaulin cover. In order to avoid an automatic opening, the door is made in such a way that it can reverse on its hinges around the standpipe, if it is slightly lifted before hand, i.e. removed from the latch pins. The pin for suspending the ballast weights is situated under the cabin floor on an arc with which the cabin is joined to the lower suspension cables.

Protection of the mast against lightning consists of a two-meter steel pipe and a contact frame. The steel pipe is welded to the ladder and angle bracket, while the contact frame is buried under the earth. The pipe and the contact frame are connected through the stringer of the ladder.

In the *PNM-3* model, ascent and descent of the observer takes less than 2 minutes, for which it is necessary to apply a force of 5 kilograms. The Altai Forest Division has developed a model of a mast in which the ascent and descent process is controlled by an electric motor. Such mechanization slightly complicates the structure of the elevator and can only be recommended in places where electric power supply is available.

Experience in the construction of fire observation towers of the *PNM-3* type has shown that the cost of the mast is half as much as that of a wooden fire tower of the pyramidal type of the same height, about 1500 rubles. However, the main advantage of such masts is that mass production in factories is possible.

The fire observation mast of the Barnaul Forest Division is basically the

PNM-3 mast with the following structural improvements: the observer's cabin ascends and descends along the ladder by means of an electric motor with a regulator and belt drive on rollers which turn the guide pulleys of the cable system. In the *PNM-3* mast, this process takes place due to the repulsion force of the ladder rungs (5–6 kg); the system of ascent and descent has no speed limit device (gears), which is characteristic of elevator systems, including that of the *PNM-3*. However, such a system of ascent and descent requires the laying of electric power lines for the mast which is not always possible under forest conditions, while the absence of trap gears is not the final solution for resolving safety problems.

Although the *PNM-2* and *PNM-3* masts are structurally successful, they do not preclude the necessity of the observer's ascent and, consequently, do not eliminate operational difficulties caused mainly by an observer's fear of height. The latter condition is one of the causes limiting the use of several already installed masts in various regions of the country. It should also be noted that comparatively frequent ascent on the mast during day light necessitates considerable physical effort on the part of the observer, and staying in the open small cabin at a height of 35–37 m, itself, entails obvious discomfort. The foregoing provides sufficient reasons for a quest for more perfect methods and means of forest fire detection, using the latest radio-transmission aids and, in particular, television as a more promising means of long distance observation. Development of the commercial television techniques enables an operator, who is present in direct proximity, to detect forest fires over wide areas.

Work on the development of special television systems for the detection of forest fires was started by firms in England, France, Canada and the United States of America in the mid sixties. As a result of experiments conducted in 1955 by the *Hancock Electronics Company,* American specialists arrived at the conclusion that by means of a television camera provided with a lens with a focal length of 500 mm, the range of detection of smoke is 20–25 km, whereas with the naked eye, the effective detection range does not exceed 2–4 km (Hastings, 1956). In the same year, the English firm, *Marconi Industrial,* developed a miniature transmitting camera weighing 2 kg for forest fire detection transmitting an image by cable to distance of nearly 6 km (*The Timberman,* 1955).

A more perfected television system was demonstrated by a Charleston Company *Rauthonon Manufacturers* (*The Timberman,* 1955). This device consisted of a series of cameras connected with a control station (for example, the forest division office) by radio link. Control of camera rotation along the horizon and transmission of the video signal is carried out by a radio channel in the shortwave band. For obtaining the location of fire area, azimuth dials are placed before the lenses in such a way that the lower part of the screen of the video-control device projects the figure indicating the direction of the fire. Entry of the same fire area in the map from two cameras gives its precise coordinates. A similar system consisting of a series of cameras with a radio channel was developed in France.

In the Soviet Union, the first experiments on the use of a television system for forest fire detection were started in 1963 at the forest Fire-Protection Department of the Leningrad Scientific Research Institute of Forestry (Artsybashev, Shtuchkov, 1965). The complete set of the experimental model of the television device for forest fire detection includes (Fig. 18): a transmission camera on a rotating table, a video-control device with remote control and a direction-finding system. The transmission camera is provided with two lenses with focal lengths of 35 and 135 mm. Operation of the tube and all the mechanisms is accomplished by remote control through a cable 200 m in length. The revolving device helps in horizontal guidance of the camera in an angle range of 330° and vertical guidance in an angle range of ±20° from the horizontal, ensuring observation of the surrounding area from a distance of 200 m from the visible skyline. The experimental model has a direction-finding system with a synchronized transmission for the angle of rotation of the camera, which helps in locating its position on the map of the forest division and obtaining the direction of the fire with an accuracy of ±1°. The transmission camera with a revolving table is mounted on the mast and is connected by a multicore cable with a video-control device which is installed in any closed building. The device is fed from AC mains with a voltage of 220 V and remains in operation during the day. At the required time, the observer adjusts the necessary screen brightness and, by operating the keys of horizontal and vertical controls, observes the panorama at first on a broad scale by using a short-focus lens and then, in case of detection of smoke at a great distance (over 5 km) by using the long-range viewer.

Field trials of the television system were carried out during the fire season in 1965–66 in the Siversk Roshchinsk forest divisions of Leningrad Province and in the Pskov Forest Division. During trials with the television set, over 20 forest fires, mainly ground fires, and 64 smoke sources of experimental fires were recorded. At the end of the trials, it was established that the detection range of forest fires was directly related to the focal length of the lens, the dimensions of the smoke cloud and the atmospheric visibility conditions. The maximum distance at which a smoke cloud of a developing forest fire of ground character, as detected by means of the television set, was 18 km. Under ideal visibility conditions and on development of the fire into a crown fire, the detection range increases.

The optimum distance at which smoke from incipient forest fires can be detected on the screen of the television set depends on the background against which it is projected. Experiments showed that when the height of the transmission camera is 15–20 m above the forest canopy and when the

Fig. 18. Television device for forest fire detection:
1—transmission camera on revolving table; *2*—videoreceiver device with control desk; *3*—fire direction finding system.

terrain is flat, all smoke clouds within a radius of 8 km will be projected against the background of the forest (dark background) and in this case, detection on the screen is not difficult. The picture is different if the background to the smoke cloud happens to be the sky. In this case, the contrast in brightness between the object and the background is comparatively insignificant, and as such the screen of the television set distinguishes smoke clouds with difficulty.

In visual observation of forest terrain, there is color contrast in addition to a contrast of brightness. The presence of contrasting colors (for example, green forest canopy and white smoke) largely determines the likelihood of early detection of forest fires by the naked eye. In a black-and-white television system, color contrasts are not possible, and the visibility range is determined only by the contrast of brightness. The maximum contrast between the object and its background will be obtained when the spectral characteristic of the photosensitive layer of the tube coincides with the maximum difference of the spectral characteristics of the object and the background. For verifying this proposition, the 'vidikon' tube in the experimental model of the television set up was replaced by an experimental tube, the maximum light-sensitivity of the photocathode layer of which is in the infrared zone of the spectrum (about 760 mm). When guiding the television camera with this tube toward distant smoke clouds, their visibility on the screen of the set, (according to the common conclusion of several observers), improves considerably. Thus, there are prospects of improving the efficiency of television systems for forest fire detection by using the principles of spectral television.

The main characteristics of a forest fire smoke cloud, which distinguish it from ordinary atmospheric clouds, are its darker color, rapid deformation, mobility and its closeness with the forest canopy. Experience shows that best results are obtained by inspecting the dynamical image on the screen, i.e. by moving the camera along the horizon. In this case, smoke clouds of forest fires having a low contrast with the background, slide along the screen due to their close connection with the forest canopy and can be comparatively easily detected from the stationary images on the screen.

Success in early detection of distant forest fires largely depends on the experience of the observer. By visually memorizing details of the terrain, which repeatedly pass before his eyes during the day on the screen of the set, an experienced observer immediately notices all low-contrast light spots, which could be smoke clouds of forest fires.

Preliminary calculations show that expenditure on the television method of forest fire observation is 1.5 times higher on the average, as compared to the usual visual method. However, the use of a television system itself greatly simplifies the process of forest fire detection making it safe and convenient and improving the technical level of the forest conservation service.

In the near future, development by the indigenous industry of new commercial television sets based on semiconductors, which are more efficient as regards light-sensitivity parameters, will offer fresh scope in the use of television for forest conservation.

Ground patrolling is organized side by side with forest observation from fire towers and masts during the fire-risk season with the aim of ensuring compliance of fire safety rules by visitors, and by people working in the forest, as well as early detection and suppression of fires. Ground patrolling is carried out mainly by foresters and temporary fire guards. However, during the period of maximum fire risk, divisional forestry technicians and fire guards of the organizations working in the forest are also pressed into service.

Ground patrolling is initially carried out in places which are most frequented by people and tourists as well as in commercial timber depots and turpentine producing areas and along roads passing through the most fire-prone areas of the forest and felling areas. The number of patrols and, the intensity of patrolling is determined by the degree of probability of fire occurrence, the value of the protected forest areas and weather conditions.

Patrolling is carried out on horseback (mainly in mountainous regions), on bicycles, motorcycles, launches motor-trolleys and motor vehicles. The length of the route and the extent of territory under control depends on the type of transport, the distance of the area from a populated area and the operational base of the patrol.

On detecting an incipient fire, a patrol should at once extinguish it by its own effort. With this aim, the patrol transport is equipped with light fire-fighting aids: shovels, knapsack, fire-extinguishers, light motor pumps, fire-extinguishing chemicals, etc. If a patrol is unable to extinguish the detected fire, it should at once inform either the forest division which has a fire-chemical station, or the nearest populated spot in order to organize suppressing of the fire by the efforts of voluntary fire brigades. For transmitting information about fire, the patrolling motorcycles, launches and automobiles are all provided with portable radio sets. In regions with a sufficiently developed network of fire towers and masts, observers detect up to 85% of the total number of fires. The remaining 15% are detected by ground patrols and the local population.

When observing forests from fire towers and masts, and while patrolling, it should be remembered that probability of fire occurrence varies sharply during the course of the day. In the morning, when the forest cover is still moist with dew, fire occurrence probability is minimum. As the sun rises, the soil cover gradually dries up and acquires the capability of ignition even with the most insignificant sources of fire (matches, cigarette butts, sparks, etc.). After midday, with decrease in temperature and increase in relative atmospheric humidity, the hygroscopic cover again lowers the forest's susceptibility to catch a fire. Thus, occurrence of fire in the course of the day is subject to the law of normal distribution. As an

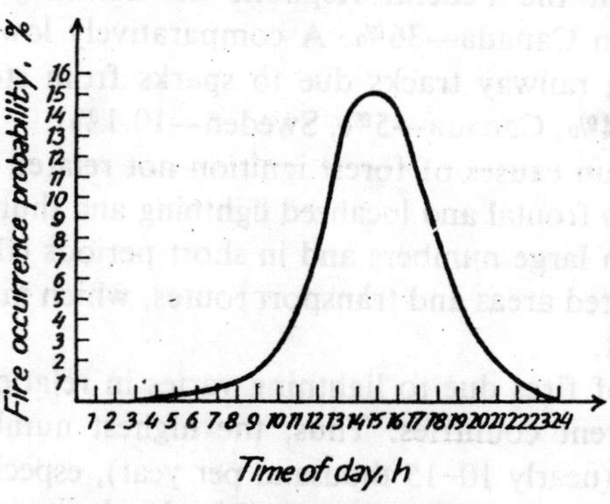

Fig. 19. Distribution of the number of fires with respect to the time of day.

example, distribution of the number of fires in relation to the time of day for conditions of the Karelian Isthmus in Leningrad Province (Fig. 19) has been given. The maximum possibility of fire occurrence is found to be between 14–15 hours in the afternoon with the minimum, in the morning and evening. Fire occurrence probability during the night is extremely low, such cases being likely only during severe drought periods accompanied by dry thunderstorms.

In forests included in the workers' recreation zone, the maximum number of fires is observed on Sundays and Mondays. This shows that the main cause of fire occurrence in these places is carelessness on the part of holiday makers and the local population. In thinly populated regions, it is extremely difficult to establish the cause of fire occurrence in forests. Therefore, reliability of the data on the distribution of the number of fires is comparatively low. In the Soviet Union, fires due to unexplained causes are nearly 35% of the total number (Antsyshkin, 1957). In European countries, it is 18–33%, in the US—17.3%, and in Canada—16–24% (Kurbatskii, 1949). The ratio of the number of fires according to established causes during 1960, 1966 and 1972 in the areas covered by air bases is given in Table 12 (Chervonnyi, 1973).

Analysis of the data in Table 12 shows that if earlier on, fires occurred in most cases, due to the fault of workers of enterprises and organizations employed in the forests, in recent times, they have been occurring more often due to the fault of the local population. Reduction in the number of cases of fire occurrence due to the fault of organizations and enterprises has been observed since 1971, when the ordinance 'Fire safety rules in the forests of the USSR' came into effect. These rules give wider powers to forest division teams in dealing with defaulters.

Careless handling of fire by people in forests is the main cause of fires in many countries of the world. For example, in Sweden it is 56%, in France—31.3%, in the Federal Republic of Germany—57.5%, in the US—46.4% and in Canada—36%. A comparatively lower percentage of fires occurs along railway tracks due to sparks from steam engines (the United States—4.4%, Canada—5%, Sweden—10.1%).

One of the main causes of forest ignition not related to the activity of man is fires due to frontal and localized lightning and thunderstorms. These as a rule, occur in large numbers and in short periods (flashes of fire) far away from inhabited areas and transport routes, which makes their control extremely difficult.

The number of fires due to lightning varies in relation to the climatic features of different countries. Thus, the highest number of such fires occurs in the US (nearly 10–15 thousand per year), especially in the region of the Rocky Mountains, where orographic clouds are often transformed into thunderclouds. In Canada, the number of fires due to lightning reaches 26% of their total number. In European countries with a high density of population (FRG, Sweden, France, Spain), the percentage of fires due to lightning varies from 1.1 to 2.8%.

In the USSR, on the whole, the number of fires due to lightning is not more than 2% of the total number of fires in the country. However, in areas under aerial protection, it reached 12% in 1966 (Chervonnyi, 1973). In serveral eastern and south-eastern regions of the country, lightning is considered one of the most important causes of fires. Thus, according to the

TABLE 12. RATIO OF NUMBER OF FIRES OCCURRING DUE TO VARIOUS CAUSES

Causes of fire	Number of forest fires, in % of the number of fires with known causes		
	1960	1966	1972
Due to fault of:			
lumbermen	38.6	10.0	5.8
expeditions	4.1	9.0	1.8
Due to agricultural fire-clearing work	10.9	16.0	9.5
Due to steam-engine sparks	15.4	9.0	1.3
Total:			
due to fault of enterprises and organisations	69.0	44.0	18.4
due to the fault of population	28.0	44.0	73.0
due to lightning	3.0	12.0	8.6

data of S.N. Uspenskii (1959) in the pine forest belts near the Irtysh river bank, the number of fires due to lightning was nearly 70% of the total fires. According to the observations of the author, 58 forest fires occurred in 1971 near the Angara River bank in the Krasnoyarsk Region during the three days of passage of a thunder trail.

The problem of detecting and controlling forest fires occurring due to thunderbolts is one of the important problems and requires a scientific and technical solution. While an overwhelming number of fires due to human failure occur around populated areas or along roads, rivers and streams, fires due to lightning are often detected in the most distant places, where their suppression becomes extremely difficult.

In the regions of Siberia and the Far East, the main cause of forest-fire occurrence in the spring is the burning which is carried out in cultivated lands in order to destroy the previous year's dry grass and enrich the soil with ash elements. Poor control of the fire often results in its spreading into the forests, especially in forest stands where herbaceous plants predominate in the cover. In logging and timber harvesting regions, fires occur mainly in the spring, when felled areas are cleared by burning, as for example, when incineration is resorted to for clearing the debris. In the middle of summer, a large number of fires occur in places where berries and mushrooms are collected.

A study of the causes and pattern of occurrence of forest fires forms a basis for working out fire preventive measures, proper organization of manpower, and methods of fire control.

Methods of Extinguishing Forest Fires

It has been observed that the following conditions are essential for ignition and the normal course of combustion process: the presence of combustibles, free access of air to the combustion zone, and an element of ignition. After ignition, the process of combustion is established, forming an interacting system, if the consumption of heat for preheating of combustibles and losses due to convection do not exceed its intake. All methods of extinguishing forest fires necessitate the elimination or dilution of one or two conditions of the combustion system, or strengthening of the resistance of the combustibles to further ignition. Therefore, all existing methods of extinguishing fires are based on the following physicochemical measures:

1. To stop the entry of air into the combustion zone.
2. To cool the combustion zone to a temperature below that of ignition.
3. To remove combustible substances from the zone of preheating and ignition.
4. To terminate the chain reaction of combustion by inhibition.
5. To increase the moisture content of combustibles to a degree at which combustion is not possible.

Practical experience has established the following methods of forest fire control, from which a particular measure, or combination of measures, can be selected: 1) beating of the flame edge of ground fires with branches, 2) covering of the flame edge with soil, 3) quenching with water or fire-extinguishing chemicals, 4) constructing, in the route of the advancing fire barrier or control belts and trenches, either manually, with mechanical tools, or by blasting and 5) annealing.

BEATING OF GROUND FIRE EDGE

Beating the flame edge of a ground fire is a primitive but very widely used method that can be employed by anyone who detects a forest fire. For beating, it is necessary to make a broom consisting of large branches of broad-leafed species which can be conveniently held with both hands. Sometimes, instead of some large branches, a young sapling with a well-developed crown is used, its lower portions having been chopped off.

The flame is beaten with sliding strokes directed toward the burnt out area. This method is more effective if the stroke falls on the extreme boundary of the fire edge, i.e. at the junction of the burning and nonburning materials. Beating actually causes separation of the flame from the spray of fuel gases, originating in the pyrolysis of the combustible material, the cooling of the combustion zone by the branches and the sweep of cold air, and the partial removal of burning particles to the burnt area behind the fire. This method is most effective when extinguishing the fire edge in light soils with a moss and lichen cover. It is difficult to use this method in a forest with a developed cover of bushes (heather, ledum, cassandra, etc.), since bushes, because of their strong and flexible stems, are shock-absorbent and the stroke of the broom does not achieve its purpose.

When extinguishing large fires, the beating method is used most often in the early morning or late evening (in northern regions even at night), when the intensity of combustion is drastically reduced, and the low atmospheric temperature lowers the fatigue limit of workers and increases the efficiency of their labor. A team of two workers, one of whom beats and the other puts out the fire edge, takes one hour of intensive work to extinguish the edge of a ground fire of low intensity over a length of nearly 600 m. However, since the reliability of this method is comparatively low when extinguishing large fires it is of secondary importance. Small fires can be extinguished by beating without involving additional technical aids.

In Canada and the United States, strips of strong rough cloth tied to a long handle with a cross beam are sometimes used in fire-fighting practice. Wet strips are dragged along the fire edge or the flame is snuffed out by them followed by covering up of the cinders. In both cases, after beating, it is essential to cover the remaining sources of ignition with soil or to localize them with a belt of exposed earth.

EXTINGUISHING FOREST FIRES WITH EARTH

The use of earth as an immediate forest-fire control measure is encouraging since, in the most inflammable types of forests growing on dry sandy soils, earth is readily available in unlimited quantities and in proximity to the fire. As compared to beating, this method can be used not only in forest stands with moss and lichen cover but also in the presence of bushes, since bushes do not burn on their own without support from below.

When the fire edge is covered with earth and the flame beaten, the entry of air into the combustion zone is checked and, simultaneously, the burning particles are cooled below the temperature of ignition. Earth is dug out with shovels from pits on the spot near the fire edge and scattered in such a manner as to beat the flame along the entire width of the edge. After extinguishing a 3–6 m long section of the edge, the next section is dealt with, again starting with removal of the upper layer of the cover and forest floor and digging out soil from the pit. After beating the flame, the fire edge is covered with an earth layer 30–40 cm in width and 2–4 cm in thickness. Individual combustion sources (for example, old stumps, deadwood) are covered with a thicker layer after removing the forest floor by digging around them. Sometimes, this method of earth covering is combined with localizing the source of the fire by a narrow belt of exposed earth.

The method of extinguishing a ground fire edge with earth has found wide application in practice, although it is very laborious and inefficient. After one hour's work, a team of two men can reliably localize only about 100–120 m of a fire edge. In recent years mechanization of this process has been considered possible with the development of a series of earth casters of varying capacities at the Leningrad Scientific Research Institute of Forestry under the guidance of N.P. Valdaiskii and A.N. Chukichev (Valdaiskii, 1970). Several models of light and heavy earth casters have been developed in the United States of America. According to their structure, all soil casters are divided into manual and tractor operated types.

One-man motorized earth caster GR-1. The main purpose of the earth caster is to extinguish the edge of forest ground fires with earth. It can be used to lay barriers and to create exposed soil belts (in the event of a backfire) and to localize the source of the fire with a ditch (Valdaiskii, Chukichev, 1969).

Earth caster *GR-1* consists of the following main parts: motor, regulator, transmission system and working device (Fig. 20). The 4.5 hp engine consists of an automatic saw, *Druzhba,* without any structural changes or alterations. For protecting the motor from shocks and for supporting the earth caster when putting it on the ground, there is a special tube frame. The motor starts with a detachable starter. The regulator serves for a quick change of soil ejection direction, depending on the movement of the worker in relation to the fire edge and a change in the magnitude of the torsional moment transmitted from the motor to the working device. The working device is a special structure which includes: a conical regulator, a thrusting gear wheel for regulating the depth of the furrow being dug, a casing for oriented earth ejection and protection of the worker from the soil spray and finally, the rotor. The rotor is a milling head with three cutting and three propeller blades of hightensile steel. In the American model of the manual earth caster of a similar structure, short link chains are adjusted in the rotor, instead of cutting and propeller blades.

During operation, the earth caster is held on the right side of the worker, so that the transmission casing lightly touches his thigh. When working, the operator should stand erect, thrust the gear wheel into the ground and, while smoothly increasing the rotations of the motor, start his movement. During movement, the thrust wheel should not be removed from the ground, since it will reduce the contra-rotating torque of the rotor. Translatory motion is carried out by the worker's efforts, but the speed of movement

Fig. 20. One-man earth caster, *GR-1*, designed by the Leningrad Scientific Research Institute of Forestry.

depends on the quality and reliability of suppression of the fire edge and maintenance of the continuity of the furrow. Depending on the forest type and vegetative conditions, movement speed varies from 0.8 to 2.5 km/h. When extinguishing a fire, the worker moves the earth caster along the fire edge at a distance of 1.2–15 m, trying to direct the soil spray under the flame base. The range and height of the soil spray are regulated by inclining the caster on either side. The quantity of ejected soil during the operation can be increased for a short duration by a smooth forced penetration of the working device into the soil by pressing the hand lever. In cases, when tactical reasons require an increase in the quantity of ejected earth and construction of a more reliable barrier belt, double or even triple gaps should be made with a small increase in the width of the belt. The maximum soil ejection range with a 45° angle of the cut-off panel is 3.5–4 m. With a depth of a furrow of 7 cm and a width on the surface of 23 cm per linear meter, nearly 14 kg of sandy soil is ejected. This quantity of soil is sufficient to beat the flame edge and cover it with a 1.5–2 cm layer. The furrow created in the process prevents further spread of the fire and reliably liquidates it.

The use of this soil caster recommended only in a dry type of forest with sandy soils (lichen-moss and whortleberry-covered pine forests) and it is advisable to work in teams of two men. The worker walking in front marks and clears the passage route, while the other extinguishes the fire edge with soil. Operation of the soil caster requires skill and stamina and, therefore, in order to avoid overexhaustion, workers should exchange roles every 30–40 minutes.

Tractor-operated earth caster. In structure, the tractor-operated soil caster, *GT-2,* is a mounted implement with an active working device of the milling type (Fig. 21). The working device is operated by the power take-off shaft of the tractor through the propeller shaft. The main units of the tractor-operated soil caster are: the body, the propeller driven shaft, regulator, transfer mechanism, safety clutches, working device, road wheel and protective casing. The working device consists of cutting blades and casting shovels suspended on hinges, which fling soil sideways to a distance of 20 m. The soil casting range can be regulated by the blades of the casing. The pivoted suspension of the cutters helps the earth caster to surmount obstacles (deadwood, thick roots, etc.). After its passage, the soil caster leaves behind a furrow with a surface width of about 100 cm and a depth of 17–20 cm.

Fig. 21. Construction of a wide barrier belt with tractor-operated earth caster, *GT-2*.

The earth caster can be operated only with tractors of three-ton capacity and above (*LKhT-55*, *T-4*, *T-140*, etc.), equipped with rear hydraulic hinge plates and having a power take-off shaft. It is intended mainly for a belt in front of a fire perimeter and also for active suppression of the edge of ground fires and of crown fires in softwood saplings. When constructing barrier belts in dry sandy soils, the speed of the earth caster assembled with the tractor *LKhT-55* is about 1.1 km/h.

EXTINGUISHING FIRES WITH WATER

Water has been known as a means of fire suppression from ancient times. It is an effective fire-extinguishing agent due to its comparatively high heat absorption capacity, thermal resistance and very high volumetric expansion coefficient during evaporation. Next to earth, it is the most widely used fire extinguishant in forest areas. Rivers, lakes, bogs, streams are found everywhere in forests and are natural reservoirs of this effective fire extinguishant. To illustrate the point, it may be noted that in Karelia alone there are over 40000 lakes, in the Kola Peninsula there are 110000 lakes. Altogether in the Northwestern Territory of the USSR, there are nearly 175000 lakes (Domnitskii, Dubrovina, Isaeva, 1971).

The fire-extinguishing properties of water are explained firstly by its cooling effect due to a high heat absorption capacity. Evaporation of one liter of water requires about 620 kcal which it can take from the combustion zone. On heating, one liter of water forms nearly 1700 cc of vapor. The vapor, continuously forming above the combustion zone lowers the concentration of oxygen in the air to the critical level (14%) and thus aids in arresting combustion. When extinguishing a fire, the effect of water sprays becomes apparent, when falling on the burning material, they separate the flame jet from fuel gases originating during pyrolysis of wood. Large jets of water, because of their high kinetic energy, break the layer of combustibles. Thus, when extinguishing fires, water is used as a cooling agent, as a means of reducing the flow of atmospheric oxygen to the combustion zone, as a means of separating the flame from the jet of fuel gases and, finally as a means of isolating burning particles of vegetation from unburnt materials.

The negative properties of water as a means of extinguishing a fire are its low heat conductivity and its high molecular surface tension (72 dyn/cm^2).

The combustion process will cease when the rate of absorption of heat by the liquid exceeds the rate of liberation of heat, the latter being used up in preheating and ignition of combustible materials. With copious amounts of water, the extinguishing process is almost instantaneous, though a large part of the water is unnecessarily wasted.

If the fire happens to be near a water reservoir, the question of the quantity of water required for its suppression, especially when motor and hand pumps are available, is not a serious one. It is another matter when water for the fire has to be fetched from a long distance or carried by any means of transport. In this case, the question of the optimum quantity of water required for a given intensity of combustion acquires very great importance.

S.M. Vonskii (1957) carried out theoretical calculations for the minimum quantity of water required for extinguishing the edge of ground fires in various types of forests with covers of different humidities, depending on wind velocity. On the basis of these calculations, he estimated the quantity of heat liberated per unit of time (for example, per second) for one running meter of fire edge. This can be found from the calorific value of combustibles taking into account the intensity of their combustion in the presence of wind. By dividing the quantity of heat liberated for one running meter of fire edge by the heat of evaporation (620) a certain theoretical value of water flow rate can be found

$$L = \frac{Q}{620} \text{ l/s.}$$

The actual consumption of water in extinguishing fire is much higher than that theoretically calculated, since a part of it inevitably falls on sections already extinguished, a part evaporates while passing through the flames and, finally, a certain quantity escapes from the combustion zone beyond the limits of the fire edge in the form of spray. To account for these unavoidable losses, S.M. Vonskii multiplied the resultant rated data by the flow coefficient, equal to 2.

The optimum water delivery rate (in liters/second) per running meter of frontal edge of the ground fire with varying flame intensity is given in Table 13. This data is confirmed by experiments conducted in forests and can be used as the initial assumptions when designing knapsack equipment for extinguishing forest fires with water.

From the data in Table 13, it may be concluded that the water flow rate for suppression is directly related to the intensity of combustion. Thus, with a seven-fold increase in wind velocity in lichen-moss pine forests, the water flow rate increases eight times. The negligible water flow rate for extinguishing a soil cover, when humidity increased to 50% and above, is worth noting.

TABLE 13. OPTIMUM WATER DELIVERY RATE (L/S) PER RUNNING METER OF THE EDGE WITH DIFFERENT WIND VELOCITY/FOREST COVERS

Wind velocity under forest canopy, m/s	Water delivery rate with humidity of lichen-moss story, %		
	up to 30	from 30 to 50	above 50
	Lichen-moss pine forest		
0.5	0.11	0.07	0.03
1.5	0.39	0.22	0.10
2.5	0.65	0.36	0.16
3.5	0.86	0.47	0.23
	Moss-covered pine forest		
0.5	0.06	0.02	—
0.5	0.12	0.03	—
	Ledum-covered pine forest		
0.5	0.17	—	—
1.5	0.44	—	—
2.5	0.74	—	—
3.5	1.02	—	—

The low heat conductivity of water is one of the properties which lowers its fire extinguishing effect. In order to accentuate this effect, efforts are made to spray water in a finely atomized form on the edge. In this case, the tiny droplets have altogether a larger heating evaporation surface. When passing through a flame, these droplets are quickly heated and transformed into steam, absorbing a large quantity of heat and localizing the fire area against the atmospheric oxygen. Thus, the warmer the water delivered to the combustion area, the quicker its transformation into steam, since steam has better fire-extinguishing properties as compared to atomized spray. This conclusion has been confirmed by laboratory experiments. However, the construction of a mobile steam-air device useful for forest conditions is technically complicated. Therefore, the practice is limited to mechanical atomization of water spray.

One of the possible versions for enhancing the fire-extinguishing properties of water can be its use in the form of a finely dispersed aqueous aerosol with droplets of dimensions from 20 to 10 mμ. This new qualitative condition of water was assumed on the basis of an idea of extinguishing forest fires with aerosol clouds, suggested by the Forest and Timber Institute of the Siberian Department of the AN SSSR. It was assumed that aqueous aerosol processed by an aerosol generator should fill up the fire area and increase the absolute atmospheric humidity and that of the combustibles to a level at which combustion is impossible. However, subsequent studies showed that in order to obtain an aerosol cloud spread over a distance of several kilometers from the site of its formation, it would be necessary to have a generator with a capacity of not less than 10 t/min even under extremely favorable conditions (10°C and relative humidity 95%) (Kutsenogii, 1970). Since a large part of aqueous aerosol will be held up by the developed surfaces of foliage and needles of trees and bushes, this would require additional increase in the power of the generator and, consequently, a huge expenditure of energy.

In practice, the use of a particular method of extinguishing fire with water depends mainly on its distance from the fire edge. When a water source and a pumping device are available, ground fires of any intensity and also crown fires, especially in young trees, can be extinguished with a powerful spray. When extinguishing ground fires in light soils with a powerful spray, the soil cover and forest floor are washed off, sometimes down to the mineral layer. Fires in forest stands with developed cover of mosses and bushes and in peat soils are also extinguished with powerful continuous sprays. However, extinguishing a fire with water, under these conditions, often creates a false sense of safety, since fire can remain concealed in hummocks, old stumps or pass into the peat horizon. Soaking these combustible components with water does not always yield the desired results. Vaporous paraffins, formed in the process of decomposition of bitumens on contact with fresh air, harden on the surface of the particles of combustibles and prevent them from being dampened. Wax coating on the leaves of ledum, cassendra, whortleberry and other bushes also do not permit them to get wet. Therefore, to enhance the fire-extinguishing properties of water, small quantities of surface-active substances, viz. wetting agents, are added to it. Suppression of fires with aqueous solutions of surface-active substances will be examined below in greater detail.

SUPPRESSION OF FIRES WITH CHEMICAL EXTINGUISHANTS

Development of the chemistry of extinguishants in the USSR began from 1930, after successful experiments in extinguishing forest fires conducted under the guidance of A.M. Simskii. These experiments established that chemical solutions can be used for extinguishing fire, as well as for creating fire-retarding barrier belts. In 1934, a study on the fire extinguishing effect of various chemicals was conducted at TsNIILKh (now Leningrad Scientific Research Institute of Forestry) under the guidance of P.P. Serebrennikov. He tabulated efficiency grades for over 20 types of chemicals and their mixtures. The works of G.A. Amosov (1958) and N.N. Krasavina (1965) are devoted to the study of the mechanism of interaction of burning wood and chemical extinguishants. These works laid the foundations of the theory and practice of suppressing forest fires with chemical extinguishants.

In recent years, significant contribution in working out new methods of extinguishing forest ground fires with chemical solutions has been made by V.G. Lorberbaum (1965, 1972). He blended several new extinguishant mixtures and emulsions and also conducted highly complex investigations on the use of highly effective foams in forest fire practice.

Suppressing forest fires with extinguishants is a complex physico-chemical process, which has not been completely studied so far. The stoppage of combustion in forest fires after the action of chemical extinguishants on them is achieved due to: 1) cooling of the combustion zone to the temperature at which combustion cannot continue; 2) isolation of the combustibles from atmospheric oxygen by a layer of gases which prevent combustion and also the formation on its surface of a hard or liquid film; 3) formation with the combustible of stable chemical compounds obstructing the entry of atmospheric oxygen; 4) inhibition (interruption) of combustion reaction in the flame phase; and 5) conversion of various substances into coal, obstructing the entry of oxygen.

Usually, during the action of chemical extinguishants on burning vegetation components, not one of the above factors, but several of them together, are effective. To estimate the effect of each of them separately is very difficult. Therefore, in forest fire control, it is usual to estimate the fire-extinguishing properties of a particular substance in comparison with water, the quenching efficiency of which is conditionally taken as a standard.

CLASSIFICATION OF CHEMICAL EXTINGUISHANTS AND THEIR CHARACTERISTICS

All extinguishants, according to their physicochemical properties, can be divided into five classes: I—solutions; II—emulsions; III—foams; IV—suspensions; V—solids (Artsybashev and Lorberbaum, 1972). Each class combines several groups (1 to 5), characterized not only by physicochemical properties, but also by an extinguishing mechanism. A classification diagram of chemical extinguishants is given subsequently (pp. 72–73).

Each group is represented by several chemical extinguishants, presently being used or planned for future use.

Solutions are the most widely used class of chemical extinguishants. Depending on the concentration of the dissolved chemical, solutions are divided into saturated concentrated and diluted categories. A saturated solution is one in which it is no longer possible under given conditions to dissolve new portions of the chemical. In such solutions, a dynamic equilibrium is established in which, as many molecules of the substance move into the solution per unit time, as are precipitated from the solution in the initial state. The concentration indicates the quantity of the solvent in the total volume of the solution. Thus, for preparing 100 liters of a 20% solution of calcium chloride, it is necessary to dissolve 20 kg of the salt in 100 liters of water.

Usually, for controlling forest fires, chemicals are used, not in pure form, but in a mixture with a substantial portion of other chemicals. The expected volume of the solution, which can be obtained on the basis of the weight of the chemical and its given concentration, is determined according to the formula

$$V = \frac{100\, G n_2}{n_1},$$

where V is the expected volume of the solution in liter; G the initial weight of the chemical in kg; n_1 the given concentration of the solution in %; and n_2 the content of the base in the chemical in %.

In the application of fire-extinguishing agents, the following categories of chemicals are used in the form of solutions: inorganic salts, surface-active substances, thickeners and smoke-generating liquids.

Concentrated and dilute (weak) solutions are used in the control of forest fires. Inorganic salts are used in the form of concentrated solution, while surface-active substances (SAS) and thickeners are used in the form of weak solutions.

Solutions of inorganic salts. For effective suppression of fire at the edge, creation of fire-retardant belts and supporting lines for starting a backfire, and aqueous solutions of inorganic salts are used: calcium and magnesium chloride, ammonium sulfate, diammonium phosphate and

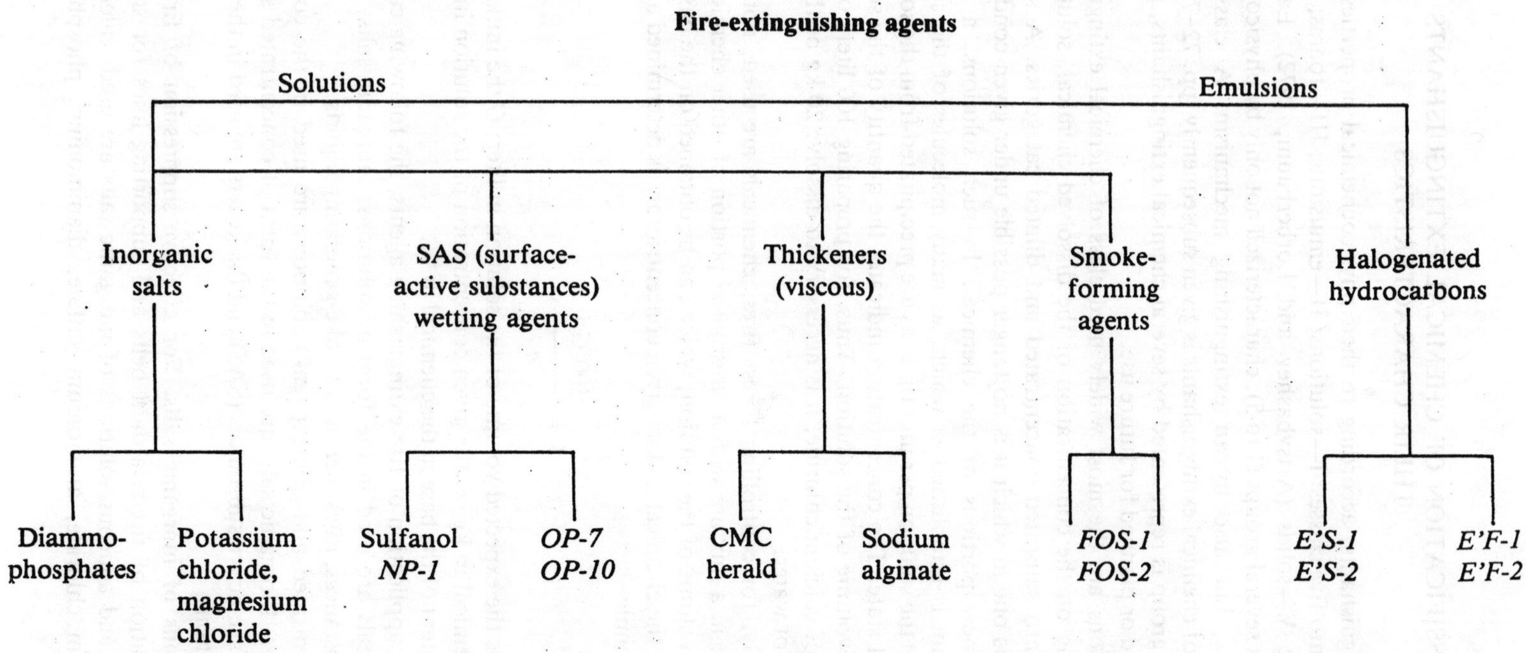
Fire-extinguishing agents
Solutions
Emulsions
Inorganic salts
SAS (surface-active substances) wetting agents
Thickeners (viscous)
Smoke-forming agents
Halogenated hydrocarbons
Diammo-phosphates
Potassium chloride, magnesium chloride
Sulfanol NP-1
OP-7 OP-10
CMC herald
Sodium alginate
FOS-1 FOS-2
E'S-1 E'S-2
E'F-1 E'F-2

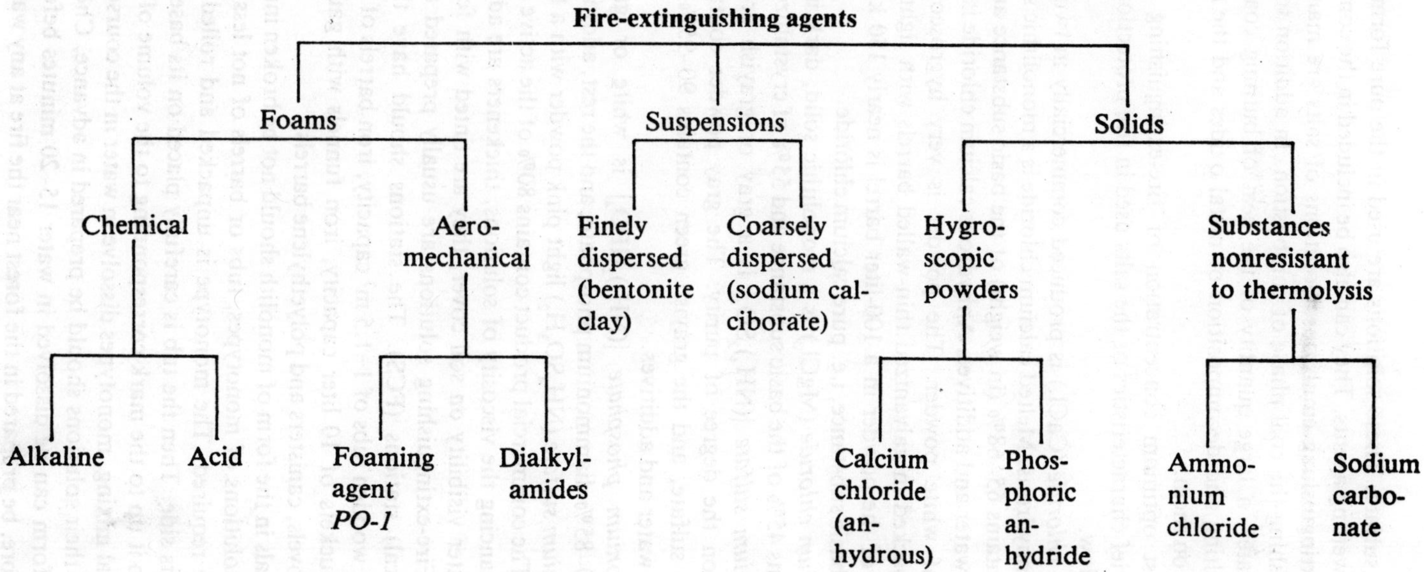
Fire-extinguishing agents
Foams
Suspensions
Solids
Chemical
Aero-mechanical
Finely dispersed (bentonite clay)
Coarsely dispersed (sodium calciborate)
Hygroscopic powders
Substances nonresistant to thermolysis
Alkaline
Acid
Foaming agent PO-1
Dialkylamides
Calcium chloride (anhydrous)
Phosphoric anhydride
Ammonium chloride
Sodium carbonate

ammonium sulfate. These solutions are used in the pure form or with the addition of wetting agents. They can also be included in the compositions of some fire-extinguishing emulsions. Solutions of salts are mainly intended for extinguishing the coal phase of combustion. In addition to the cooling action of water, a large quantity of the heat of burning coal is used in, heating, melting and decomposition of metal oxides and the acid radicals which settle on them.

The most optimum concentration of fire-extinguishing solutions is 15–20%. Brief characteristics of the salts used in fire protection operations are given below.

Calcium chloride ($CaCl_2$) is produced commercially in two forms, viz. liquid and anhydrous. Melted calcium chloride is a monolithic solid, gray in color. It contains 65–68% (in weight) of the basic substance and the rest is crystallized water and additives. Anhydrous calcium chloride is produced in the form of white powder. The powder is very hygroscopic, and is, therefore, packed in galvanized thin-walled barrels with tight fitting lids. The weight of the powder in a 100-liter barrel is nearly 150 kg containing 90% of the basic substance, i.e. pure calcium chloride.

Magnesium chloride ($MgCl_2$) is a monolithic solid, dark in color. The solid contains 45% of the basic substance and 55% of crystallized water.

Ammonium sulfate [$(NH_4)_2SO_4$] is a gray or grayish green powder, depending on the degree of purity. The gray powder contains 99% of ammonium sulfate, and the grayish green contains 90–95%. The rest is crystallized water and additives.

Diammonium phosphate [$(NH_4)_2HPO_4$] is white or gray powder. Contains 80–85% diammonium phosphate, and the rest, additives.

Ammonium sulfate ($NH_4SO_3H_2$) light pink powder with a faint smell of ammonia. The commercial product contains 80% of the active substance.

For enhancing the viscosity of solutions, thickeners are added to them, and for better visibility on soil cover, they are tinted with ferric oxide or radanine. Fire-extinguishing solutions are usually prepared at forest fire (fire-chemical) stations (FCS). The stations should have the following equipment: wooden tubs of 1–1.5 m^3 capacity, iron barrels of 200–300 liter capacity, buckets of 10 liter capacity, iron funnels with gauze, oar type wooden shovels, canisters and polyethylene barrels.

Chemicals in the form of monolith should not be broken into pieces. For preparing solutions of monotypes, tubs or barrels of not less than 1 m^3 in volume are required. The monotype is unpacked and rolled into the tub tipped on its side. Then the tub is carefully placed on its base and water is poured into it up to the mark corresponding to the volume of the solution. In periodical mixing, monotypes dissolve in water in the course of 24 hours. Therefore, their solutions should be prepared in advance. Chemicals in the powdered form can be dissolved in water 15–20 minutes before use. They can, therefore, be prepared in the forest near the fire at any water source.

Solutions of surface-active substances. Experiments conducted at the Leningrad Scientific Research Institute of Forestry showed that the humidity of dry forest floors increases slightly, even with a dose of 5–10 liters of water per square meter, while in extinguishing peat fires with water, only 20–30% of the fluid is utilized in liquidating combustion. That is why even after the suppression of peat or constant ground fires with water, they can revive.

Effective suppression of forest fires with fire-extinguishing agents is promoted by surface-active substances, viz. wetting agents. It is advisable to introduce surface-active substances up to a certain limit in fire-extinguishing fluids for lowering their surface tension. Excess of surface-active substances above the optimum level in practice does not lead to a further lowering of the surface tension of water or that of solutions of fire extinguishing chemicals (Fig. 22). The reduction of the surface tension of liquid depends on the chemical composition of the wetting agent and its concentration. The limiting (minimum) value of tension which can be achieved by introducing 2% of the wetting agent into the fluid, is 32 dyn/cm^{-1}.

In controlling forest fires, the following surface-active substances are used: sulfanol *NP-1,* wetting agent NB (*neckal'*), the boosting agents *OP-7* or *OP-10* and the cleansing agent *Progress.*

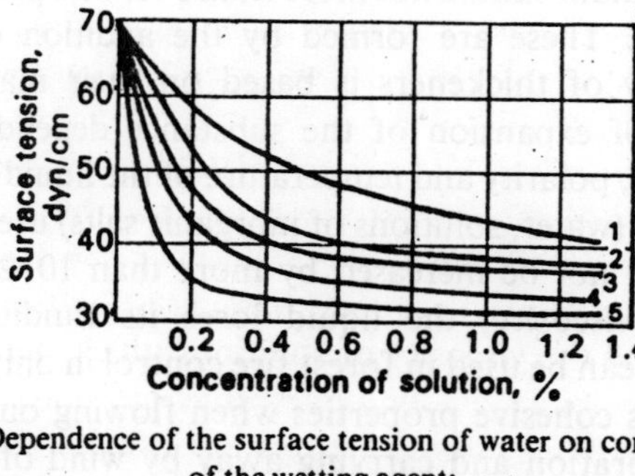

Fig. 22. Dependence of the surface tension of water on concentration of the wetting agent:
1—*Progress; 2*—*OP-7, OP-10; 3*—RAS; *4*—wetting agent, *5*—sulfanol.

Sulfanol NP-1 is a mixture of sodium salts of alkyl-benzo-sulfonic acids with alkyl residues. It is produced in the form of powder, cream to light yellow in color, containing about 50% of the active substance. The content of nonsulfurized compounds is not more than 3%, and of water, not more than 10%. The pH value of the aqueous solution varies from 7 to 8.

Wetting agent NB (neckal') is a butyl-naphthalene-1 sodium salt of sulfonic acid. The product is obtained by alkalization of naphthalene with butanol and sulfurization with monohydric salt of sulfuric acid followed by neutralization with sodium hydroxide. In appearance, the wetting agent—NB is a paste, of a gray to brown color. The content of substances insoluble in water is not more than 0.3%, and of dry matter—not less than 60%. The specific gravity is 1.15; and the pH of a 1% solution is 9–10.

Booster substance OP-7 is a mixture of polyethylene-glycol ethers of mono- and dialkylphenols. *OP-7* is produced in accordance with the Soviet standard GOST 8433–57. It is an oil-like liquid or paste of a light yellow to light brown color. It has a specific gravity between 1.06 and 1.08. Between 10° and 15° it assumes a paste-like form. The pH value of a 1% solution lies between 6 and 8. *OP-7* solution is stable to inorganic salts, in particular to calcium, magnesium and aluminum salts, and also to acids and alkalies.

Booster substance OP-10 is also a mixture of polyethyleneglycol ethers of mono- and dialkylphenols. *OP-10* is manufactured in accordance with the Soviet standard GOST 8433–57. It is an oily liquid or a paste of light yellow or light brown color. It has moisture content of not over 0.5%. At 20°C it has a specific gravity between 1.06 and 1.08.

Cleansing agent Progress is a mixture of alkylsulfates of sodium with a base of sulfuric acid ethers and secondary alcohols, the alkyl residue of which contains 8–18 atoms of carbon. It is a liquid yellow to brown in color, without sediment or exogenous inclusions. The content of active substance in the finished product is not less than 20%, proportion of nonsaponifying components in relation to the active substance not more than 1.6%, and of sodium sulfate not more than 5%. The pH is 8–9.

Viscous solutions. These are formed by the addition of thickeners to liquid. This property of thickeners is based on their marked expanding power. The degree of expansion of the substance depends mainly on its chemical composition, polarity and temperature of the liquid.

Viscosity of fluids (water, solutions of inorganic salts) used for the control of forest fires should not be increased by more than 10–20 times. With a further increase in viscosity, the liquid loses its fluidity, acquires the property of jelly and can be used in forest fire control in only specific cases.

On account of its cohesive properties when flowing out of an aircraft, disintegration, evaporation and carrying away by wind of viscous fluid is reduced. Without the thickener, approximately 40% of the liquid (water or water with surface-active substance) evaporates.

Viscous fluid adheres better to the combustibles and, therefore, is effective in extinguishing crown fires, which are the most difficult to suppress. A nonviscous fluid rolls off the surface of the combustible material before adequately wetting it.

After treatment with viscous water, components of the forest soil cover do not ignite during fire-risk weather for two-to-three hours. However, after this period, viscous water loses its efficiency. For prolonging its effect, 15–20% of calcium chloride is added to viscous water. The resultant mixture of the thickener and the salt (gel of the solution), unlike the viscous water, evaporates from forest combustibles more slowly, and after they dry up, a noncombustible film forms on the surface of vegetation, protecting it from ignition.

In wetting power, viscous compositions possess the same properties as ordinary water. The time of complete wetting of the dry combustibles is a few hours. For increasing the wetting power of thickeners, surface-active substances are added to them. For example, addition of 0.3% sulfanol *NP-1* to a viscous solution reduces its surface tension by half. In a liquid with a surface-active substance, the rate of expansion of thickeners increases by two times. After treatment of lichens, green mosses or nonconiferous branches with water plus the addition of thickeners and a wetting agent, the quantity of fluid retained by these materials increases by 1.5–2.5 times, as compared to that with ordinary or viscous water. Due to the introduction of surface-active substance in the viscous fluid, the process of drying up of forest combustibles takes twice as long.

The following thickeners are considered most effective:

Methyl cellulose (MC) and *carboxyl-methyl cellulose* (CMC). These are high molecular-weight compounds, which in appearance are similar to white unraveled pieces of woolen thread. They expand intensively in water and in solutions of antipyrens. The optimum concentration of these thickeners in water is 1%. The process of expansion in water with a wetting agent takes not more than 15 minutes.

Sodium algynate or sodium salt of algynic acid is obtained from seaweed and comprises dark colored flakes. The process of expansion in water lasts nearly 24 hours. The optimum concentration for a viscous solution is about 2%.

Polyacrylamide consists of granules of light pink color. As compared to the above-mentioned thickeners, only 0.5% is required to be added to water. Its expansion rate in water is similar to that of sodium algynate.

Concentrated viscous mixtures of thickeners should be prepared in advance, so that later they can be diluted with water on the spot to the required consistency. Concentrated solutions of thickeners can be transported to a forest in any container recommended for the transport of fluid forest-fire chemicals.

Smoke-forming solutions. These are chemical compounds, which when sprayed, produce stable aerosol (smoke or fog) consisting of tiny particles suspended in the atmosphere. The particles may be liquid or solid, and a simultaneous presence of both is possible. Aerosol is formed as a result of a hydrolysis reaction of certain substances with the water vapor present in the atmosphere.

Smoke-forming solutions can be used mainly for the active suppression of fire at the edge of forest ground fires of weak, medium and severe intensities. The most suitable smoke agents for extinguishing fire are oxychloride of phosphorus and tetrachlorated compounds of titanium, tin and silicon. Their hydrolysis occurs at temperatures 300–400°C according to their formulas.

$$2POCl_3 + 3H_2O \rightleftharpoons 6HCl + P_2O_5;$$
$$TiCl_4 + 2H_2O \rightleftharpoons 4HCl + TiO_2;$$
$$SnCl_4 + 2H_2O \rightleftharpoons 4HCl + SnO_2;$$
$$SiCl_4 + 2H_2O \rightleftharpoons 4HCl + SiO_2;$$

Unlike solutions of salts, wetting agents, halogenated hydrocarbons, etc., these extinguishants have the ability of simultaneously extinguishing the flame and coal zones of combustion. The effect is achieved with low concentrations of their vapor in the atmosphere (Table 14). For example, it is sufficient to introduce only 2% phosphorus oxychloride into the air of the combustion zone to bring about immediate liquidation of fire.

TABLE 14. FIRE-EXTINGUISHING EFFICIENCY OF VARIOUS CHEMICALS

Extinguishants	Dilatational fire-extinguishing concentration of vapors in air, %	Efficiency
Water (steam)	32.0	1.0
Silicon tetrachloride	15.0	2.1
Tin tetrachloride	3.2	10.0
Titanium tetrachloride	2.3	14.0
Phosphorus oxychloride	3.0	16.0

In spite of the foregoing, the use of $POCl_3$ for extinguishing fires is very difficult due to the irritant effect of its hydrolysis products on human respiratory organs. Besides, $POCl_3$ intensively damages some metals and alloys used in the manufacture of forest fire equipment. For lowering the rate of hydrolysis of $POCl_3$ and reducing its irritant effect on human respiratory organs, it is recommended to dissolve it in other liquids which are mutually soluble and do not hydrolyze in air. Carbon tetrachloride (CCl_4) and ethyl bromide (C_2H_5Br) have been tried as solvents. The best quenching effect was obtained, when $POCl_3$ was mixed with these substances in the proportion of 1:1. The resultant mixtures are called smoke-forming solutions *FOS-1* and *FOS-2*. Depending on the content of $POCl_3$, they have different boiling temperatures. With any $POCl_3$ content, the boiling temperature of the solution is lower than that of the pure product. Thus, the boiling temperature of $POCl_3$ (105°C for the pure product) in FOS mixture is lowered to 50°C. It is natural that when introduced into the combustion zone, such a solution changes from a condensed to a vaporous state and creates the required gaseous fire-extinguishing concentration. Hence, it can be assumed that the similarity in fire-extinguishing properties of pure $POCl_3$ and a mixture of FOS are precisely due to this behavior. In fire-extinguishing properties, the smoke-forming solutions of *FOS-1* and *FOS-2* are 5–6 times more effective than water.

Mixtures of FOS should be prepared just before use, since oxidation—reduction in reactions takes place when the prepared mixture is stored for more than one hour. This leads to a liberation of free chlorine or bromine and a lowering of the efficiency of the mixture. Therefore, the apparatus for FOS has two reservoirs: one for $POCl_3$ and the other for CCl_4 or C_2H_5Br.

Corrosive properties of *FOS-1* and *FOS-2* are the same as of $POCl_3$. Equipment for their use should be manufactured from brass, duralumin, fluoroplastic, fiberglass plastic or polyethylene at low pressure. In addition to these, steel-3 covered with ethynol lacquer can be used. For protecting human respiratory organs from the effect of vapors of $POCl_3$ and its hydrolysis products, gas masks should be used during preparatory work and directly before extinguishing the forest fire.

Emulsions. Investigation has established that fires can be effectively quenched by using a mixture, which includes the solution of inorganic salts and halogenated hydrocarbons. The salt solution thoroughly extinguishes coals, while halogenated hydrocarbons extinguish the flame. These liquids are not mutually soluble, i.e. one of them is dispersed in the other in the form of minute droplets. Therefore, emulsions can be formed using any of these liquids as a base. Emulsions are used mainly for active suppression of the edge of forest ground fires of weak, medium and severe intensities.

Emulsions can exist only during vigorous shaking or mixing. On stopping this physical action, the liquid at once separates into two independent layers. For obtaining stable emulsions, a third component, viz. an emulsifier, is added.

Carbon tetrachloride (CCl_4), ethyl bromide (C_2H_5Br) and Freon 114 B2 are used as gaseous extinguishants in fire-extinguishing emulsions. Emulsions *E'S-1, E'S-2, E'F-1* and *E'F-2* are developed on a halogenated hydrocarbon base. The composition of *E'S-1* is: carbon tetrachloride—30%, a 20% solution of calcium chloride—69.3%, and *OP-7* or *OP-10* (emulsifier)—0.5%. In the *E'S-2* emulsion, the solution of calcium chloride is replaced by a 20% solution of magnesium chloride in the same proportion. The extinguishing power of both the solutions is 2.5–3 times higher than that of water.

The basic component of the emulsion is carbon tetrachloride, which is a noncombustible liquid with a characteristic odor, a specific gravity of 1.6 and a boiling temperature of +77°C. Depending on its purity, the liquid may be colorless, or yellowish. Carbon tetrachloride does not damage clothes or harm the skin of hands, completely dissolves mineral oils, rubber, cement, etc. and is mainly used as a solvent. For obtaining 100 liters of emulsion *E'S-1* (or *E'S-2*) the following quantities of chemicals and water are required:

Carbon tetrachloride, liters	30
Calcium (or magnesium) chloride, kg	13.8

Emulsifier *T-2*, kg	0.2
OP-7, OP-10, kg	0.5
Water, liters	69.3

When preparing stable emulsions, the sequence of mixing the components should be strictly observed: CCl_4 and the emulsifier are continuously stirred while adding the salt solution; stirring is stopped after the formation of a homogeneous emulsion of a milky color. Complete preparation of a 100 liter emulsion usually takes 10–15 minutes, provided that the salt solution is prepared in advance.

Emulsion *E'F-1* includes the following substances: Freon 114 B2—5%, ethyl bromide—5%, proxanol 226 or *OP-4*—0.5%, water—89.5%. In emulsion *E'F-2* water is replaced by a 15% solution of ammonium diphosphate or monophosphate in the same proportion. In extinguishing capacity, *E'F-1* is three times more effective than water while *E'F-2* is five times more effective.

Freon 114 B2, or tetrafluorobromethane ($C_2Br_2F_4$), is a light odorless liquid, with a boiling temperature of +47.5°C. It dissolves in organic fluids in any proportion.

Ethyl bromide (GOST 2658–56) (C_2H_5Br) is a colorless liquid with a distinct smell and boiling temperature of +37 to 40°C. It dissolves well in organic solvents.

Proxanol 226 is a viscous light yellow paste with a mild aroma. It has a high stabilizing power in emulsions consisting of water or solutions of inorganic salts and halogenated hydrocarbons. It is highly soluble in ethyl bromide and Freon 114 B2. In water and aqueous solutions of inorganic salts, dissolution proceeds very slowly.

OP-4 is a thick viscous fluid of light brown color and dissolves well in organic solvents, including Freon 114 B2 and ethyl bromide.

For obtaining 100 liters of emulsion *E'F-2* the following quantities of chemicals and water are required:

Freon 114 B2, liters	5
Ethyl bromide, liters	5
Ammonium diphosphate, in kg	13.5
Proxanol emulsifier 226, or *OP-4*, liters	0.5
Water, liters	89.5

Emulsion *E'F-2* is prepared in the following manner: a 15% solution of ammonium diphosphate in one container and Freon 114 B2 in another container (polyethylene canister or barrel) are diluted with ethyl bromide in the ratio 1 : 1. For example, for preparing 100 liters of emulsion of *E'F-1* or *E'F-2*, 5 liters of Freon and 5 liters of ethyl bromide should be taken, 0.5 kg of proxanol 226 or *OP-4* should then be added to this mixture, which dissolves completely in five-to-seven minutes. After preparing the ammonium diphosphate solution, the fire extinguishing emulsion can be obtained readily.

In order to prepare 100 liters of stable emulsion, 89.5 liters of ammonium diphosphate solution should be gradually added to 10.5 liters of halogenated hydrocarbon and the emulsifier, stirring vigorously and continuously. Mixing is stopped after obtaining a homogeneous emulsion of milky color. Prepared emulsions do not deteriorate for 30–40 minutes.

For preparing an emulsion of *E'F-1*, same quantity of water is added to the mixture of organic fluids with the emulsifier, instead of the salt solution. The basic content of *E'F-1* is water. Therefore, this emulsion should be prepared directly in the forest, where a water source is available near the fire. Under these conditions, 10.5 liters of the concentrate are required for every 100 liters of emulsion. Preparation of 100 liters of emulsion from the concentrate does not take more than 15 minutes.

Foams. In the control of forest fires, foams are used mainly for building up a fire-retardant belt and supporting line along the fire, although they can be successfully used even for active suppression of the fire edge. Foam is a mass of fine bubbles filled with air or some other gas, separated from one another by a film of fluid. The size of the bubbles and their shape depend on the method of preparation of the foam and the chemical composition and concentration of the foaming agent. Foam is obtained with the spherical diameter of bubbles from 0.1 to 50 mm and more.

Of all the physicochemical properties (stability, viscosity, heat conductivity) of foam, the most important for extinguishing forest fires is its stability, i.e. the capacity to keep in form without much disintegration. The stability of foam is conditionally measured by the time (minutes) during which its initial volume is reduced by 20% (Table 15).

When extinguishing forest fires, foam reduces the flame, partly or completely isolates the flow of oxygen to the area of combustion and cools

TABLE 15. EXPANSION RATIO AND STABILITY OF FOAM DEPENDING ON THE FOAMING AGENT

Foaming agent	Concentration in water, in %	Expansion ratio	Stability in minutes
PO-1 (control)	6	60–100	15
Ammonium salt of primary and secondary alkyl sulfates (content of basic substance 29.3%)	2	130	60
Triethanol-amino salt of lauryl sulfate	1.5–2	340	120–180
Diethanolamide (C_{10}–C_{13})	2.0	180	60
Sulfate of oxyethylated fatty spirits (content of basic substance 30%)	2.0	380	60

the burning object to below its ignition temperature which, as a result, checks further combustion.

In order to enhance the fire-extinguishing properties of foam, the gas bubbles are filled, not with air, but with some fire-extinguishing chemicals in the gaseous state, particularly Freons (trifluoromethyl bromide, difluoromethyl bromide, etc.). By doing this, the fire-extinguishing properties of foam are enhanced by 5–10 times.

At present, chemical and aeromechanical foams are used for extinguishing fires. The formation of chemical foam is based on the interaction of two reagents while mixing alkaline and acid portions of the extinguishant charge. This reaction results in the formation of foam and carbon dioxide which creates considerable pressure in the container and expels the foam through the opening in the form of a long spray.

Alkaline and acid compounds are used for the extinguishant charge. The alkaline compound weighing 450–540 g consists of a powdery mixture of sodium bicarbonate and licorice extract, although instead of the licorice extract, other stabilizers may sometimes be used. The acid compound weighing 330–440 g is a powdery mixture of ferric sulfate and sulfuric acid. The extinguishing power of chemical foam is enhanced by introducing 6.4–8.4% of Ω-sulfonated phenol-formaldehyde resin in the alkaline portion.

Aeromechanical foam is a mixture of air, water and a foaming agent. Specially manufactured commercial fluids, marketed under the trade mark *PO-1* and *PO-6*, are used as a foaming agent. The composition of foaming agent *PO-1* includes: sulfuric acid—84 ± 3%; bone cement—4.5 ± 1%; ethanol (raw product) or ethylene glycol—11 ± 1%; and caustic soda, in quantities sufficient for neutralization. The specific gravity of the foaming agent is not less than 1.1; the freezing temperature is 8°C, the reaction of the agent is neutral or slightly alkaline, and the optimum concentration in water is 4–6%. Sulfanols *NP-1, NP-5, B,* paraffin sulfonate, purified alkyl sulfates of secondary alcohols, nonionic substances, etc. can be used as foaming agents. However, foams prepared on a base of the above-mentioned substances are not stable for more than 13 minutes.

In practice, there are the so-called solid foams (foam-type insulating materials), distinguished from fluid-based foams by a better tenacity of their walls dividing the bubbles. Their breakdown is possible under the effect of heat, ultraviolet rays and mechanical stress. Urea-formaldehyde resin (*MF-17* and a 'binding agent'), a foaming agent, water and acid serve as the material for obtaining solid foam.

Solid foams are not used directly for extinguishing fire. However, they can be used for fire-retarding barriers in forests in mountainous regions or, areas of dense growth i.e. in places where it is not possible to use soil working implements.

Solutions of foaming agents are prepared just before extinguishing a fire. Foam is obtained from the solution by means of foam generators. Manufacturing plants supply foaming agents in iron containers of 100 or 200-liter capacity. For facility of transport, the chemicals are poured out into 20-liter polyethylene or iron canisters. Foaming agents should be stored in factory packing in a warm place at a temperature not lower than + 5°C.

Suspensions. These are systems consisting of a fluid dispersion medium with solid particles suspended in it. They are distinguished by the amount of dispersion of the solid elements and by the degree of dispersity, i.e. by the concentration and the size of the suspended particles. According to the degree of disintegration of the solid elements phase, suspensions are conditionally divided into coarse-grained (size of the particles 100μ and above), fine grained (500–100 mμ) and turbid (less than 50 mμ), which do not precipitate under the effect of gravitational force. Suspensions are usually obtained by mixing water and solid substances diluted to the desired degree of dispersion. For enhancing the stability of suspensions, thickeners are added to them in small concentrations which reduce the settling rate of the particles.

Aqueous suspensions of bentonite clay and sodium calciborate are used in forest fire control. Bentonite clay is a finely dispersed powder of light gray color consisting of Si-82%, Al_2O_3—2%, Fe_2O_3—12%, CaO, MgO—3%, and K_2O—1%. Bentonite is added to water in quantities of 100 kg per 100 liters. During the mixing process, it absorbs water and increases in volume by 12–15 times. As a result, a finely dispersed viscous suspension is formed that does not break up for several hours, or even days, after preparation. On prolonged storage, the suspension acquires the consistency of a thick fluid, but after stirring, regains its initial properties, and can then be used.

Components of forest soil cover moistened with bentonite suspension do not burn for 2–3 hours. It is therefore used mainly as an effective fire-retardant.

Sodium-calcium borate (boric acid salt) consists of calcium borate ($Ca_2B_6O_{10}\cdot H_2O_5$) and sodium calcium borate ($NaCaB_5O_9\cdot 8H_2O$). It is a finely ground, whitish pink powder. On mixing with water in a state of rest, a comparatively thick, stable suspension is obtained. For preparing an effective mixture, 50–55 kg of powder should be added to 100 liters of water. The specific gravity of the suspension should be 1.4 g/cm^3 in order that it penetrate effectively through foliage. Forest soil cover treated with suspension serves as an effective barrier against fire spread for a longer duration than cover treated with plain water or water containing a wetting agent. Suspension retards the evaporation of the mixture from the combustibles, and after its evaporation, forms a white, incrusted, noncombustible film over it. The disadvantage of the suspension is a high concentration of the chemical in water and its corrosive properties affecting fire-fighting equipment.

Suspensions are mainly used for controlling forest fires from the air by constructing fire-retarding belts and for active suppression of the flame at the fire edge. For their preparation and use, special equipment has been developed in the United States of America and in Canada, taking into account the aircraft characteristics.

Solid extinguishants. Solid extinguishants are used for fire control in the form of finely dispersed powders of certain chemicals. Use of these substances in a finely dispersed state is necessary because of the fact, that with an increase in active surface, surface energy is also simultaneously increased. This increases the chemical activity of the substance and accelerates physical and physicochemical processes. Those reactions, which proceed rather slowly with substances in a coarsely dispersed state, are found to proceed very rapidly in a finely dispersed state. This explains the explosive property of finely ground sugar, flour, starch, coal and other substances.

Depending on the physicochemical properties of the original substances, the fire-extinguishing effect of powders may be manifested in any combination of the following effects: cooling of the flame, dilution of fuel gases by products of thermolysis of the substance, anticatalytic and antioxidant effect, hygroscopicity, and the formation of stable chemical compounds with a reaction on combustibles. The cooling of the flame takes place due to the removal of heat by the powders from the combustion zone for their own heating, melting, boiling, evaporation and decomposition. The lower the temperature of the substance at which these changes take place, the higher its fire-extinguishing property. During thermolysis, some substances decompose with the formation of a product in gaseous form. This results in a partial dilution of the fuel mixture, although this effect does not substantially influence retardation of the combustion process. The active ingredients of the combustibles which possess high energy reserves, transfer a part of the energy to the powder on contact with the powder surface and are themselves transformed into inactive ingredients. This results in breaking the chain reaction of combustion. Hygroscopic properties of powders are manifested in the following manner. The powder applied on the combustibles absorbs atmospheric moisture intensively and thus simultaneously moistens the combustible. When some powders are used, an acid is formed on the combustible surface as a result of thermolysis or their saturation with moisture, which then chemically combines with the combustibles.

The dispersion capacity of powders has a substantial effect on their fire-extinguishing capability: the finer the powder, the higher its fire-extinguishing effect. The powders, used in extinguishing fires, have particles with diameters of 5–12 μ. The dispersion capacity may be expressed by their total surface area. For effective suppression of fire, the specific surface of the powders should be in the range of 3,200 to 11,500 cm^2/g.

The following chemicals in powdered form have been tried and are recommended under forest conditions: ammonium diphosphate, phosphoric anhydride, calcium chloride and soda. Diammonium phosphate decomposes on thermolysis into ammonia and phosphoric acid. Ammonia dilutes the fuel gases, while phosphoric acid interacts with combustibles. Phosphoric anhydride intensely absorbs moisture and enters into a chemical reaction with it, forming phosphoric acid. Calcium chloride (anhydrous) attracts atmospheric moisture. In this case, the quantity of this moisture is several times more than that in the dry powder. Soda (sodium bicarbonate) decomposes in flame, releasing carbon dioxide which dilutes the fuel gases. Hexachloroethane or stearates of copper or iron are added to the powders for reducing caking and adhesion.

Substances with hygroscopic properties should be used for creating fire-retarding belts in forests. Such substances are P_2O_5 and $CaCl_2$ (anhydrous). The most suitable for active fire suppression is a composition containing 45% diammonium phosphate, 45% soda and 10% hexachloroethane. This composition is effective in extinguishing burning coal.

Fire-extinguishing chemicals are used for setting up fire-retarding barrier belts, supporting lines for the start of backfire, active suppression of fire at the edge and also for liquidating the remaining combustion areas at a fire edge after putting out the flame phase of combustion. Fire can be stopped by means of fire-retarding belts, but after that it is essential to work for its complete liquidation. Arrest of spreading fire, especially its frontal part, is the primary task. Barrier belts are created by means of solutions of inorganic salts with surface-active substances (SAS), foams, viscous solutions and hygroscopic powders.

An effective method of extinguishing fire is the direct action of the extinguishant on the flame of the fire edge. Depending on the intensity of flame, concentrated or dispersed spray is used. Conditions permitting, it is advisable to extinguish fire with dispersed spray, since the extinguishing efficiency of any fluid is proportional to its degree of dispersion. The fluid spray is delivered at the flame base along the fire line, ensuring the suppression of a longer section of fire edge. Extinguishing emulsions, *E'S* and *E'F*, solutions of inorganic salts with SAS and solutions of surface-active substances are used for active suppression of fire.

Suppression of a small fire should begin on those tactical elements of its edge where it is essential to liquidate the flame on priority. Conditions permitting, suppression of the flame should start at the frontal edge, since it has the highest flame intensity and spread speed. The flames at the flank and rear should be extinguished later. With a sufficient number of workers and a small fire area, suppression of fire should start from the rear and subsequently move forward along the flank toward the frontal edge, reducing it to a wedge. With such an extinguishing pattern, quick liquidation of the fire is ensured. Besides, the workers, moving along the direction of the wind, do not find themselves in the smoke blanket zone.

There are also other extinguishing patterns: simultaneous extinguishing of the flame at the flanks and frontal edges, extinguishing of only the flanks, etc. However, in each individual case, a fire extinguishing pattern is evolved from analysis of its development, taking into account the relief of the terrain, forest vegetation and meteorological conditions.

The fire edge is extinguished with extinguishant fluids by teams, each consisting of two workers equipped with knapsack apparatus. The first liquidates the flame, whereas the other completely extinguishes the combustion areas at the edges. Both the workers engaged in extinguishing the fire should be provided with the same number of helpers for reloading the sprayers from the service station and burning them to the fire edge. The auxiliary workers should also be entrusted with verifying the results, and, in case of revival of the fire, taking urgent steps leading to its liquidation. With such labor organization, the workers' brigade should be divided into two groups: principal and auxiliary. The principal group should include workers with experience in forest fire fighting. In case of the same qualification of workers, they should take turns, since as a rule, fighting flames itself requires a great effort and energy. In all cases, it is essential to ensure continuous suppression, which is the basic requirement for fire liquidation in a short time over a wide area. The leader of the group should keep a close watch on the extinguished edge, since revival of the fire or transfer of the flame through the barrier belt can render influence to all the manpower and means employed.

The tentative requirements of chemicals for fire extinguishing can be

TABLE 16. CONSUMPTION OF EXTINGUISHANTS PER RUNNING METER OF GROUND FIRE DEPENDING ON INTENSITY OF COMBUSTION

Chemicals	Consumption of chemical in ml with combustion intensity		
	weak	medium	severe
Emulsions:			
E'F-1	From 40 to 55,	From 70 to 110	From 110 to 160
E'F-2	,, 35 ,, 45	,, 60 ,, 80	,, 90 ,, 120
E'S-1	,, 100 ,, 120	,, 120 ,, 180	,, 180 ,, 250
E'S-2			
20% solution of	,, 130 ,, 150	,, 150 ,, 220	,, 220 ,, 290
$CaCl_2$ + SAS,			
$NH_4SO_3H_2$ + SAS,			
$NH_4H_2PO_4$ + SAS	,, 160 ,, 190	,, 280 ,, 330	,, 300 ,, 500
Water + SAS			

determined on the basis of the relative consumption rate per unit length of the fire edge, depending on the intensity of combustion of the flame (Table 16). The table gives the consumption in the liquidation of flame as well as on complete suppression of the fire. The best cost-effective result in the use of chemicals is achieved when the flame is extinguished with emulsions and the quenching of burning coal is completed with aqueous solutions of SAS or inorganic salts.

The speed of extinguishing fire with chemicals depends on the efficiency of the chemicals used, the intensity of the flame, the specifications and the operational characteristics of the sprayers, and the organization of labor. When using solutions of inorganic salts or wetting agents, the average rate of extinguishing the edge of a ground fire of medium intensity is 20–30 m/min, while with the use of emulsions, this rate is 30–50 m/min. After extinguishing the fire, it is essential (depending on weather and forest vegetative conditions) to periodically examine the extinguished edge.

Machines and Devices for Extinguishing Forest Fires with Water and Fire-extinguishing Chemicals

Machines and devices for extinguishing forest fires with water can be divided into three groups according to their structure and specific purpose: 1—knapsack fire extinguishers and sprayers; 2—motor pumps and hand pumps; and 3—fire tankers (see diagram on page 89).

KNAPSACK FIRE EXTINGUISHERS AND SPRAYERS

Knapsack sprayers for extinguishing forest fires first appeared in the nineteen thirties. These were based on knapsack devices with spray guns and piston pumps and were intended for chemical control of harmful insects and diseases in orchards and kitchen gardens.

The utility of knapsack sprayers is that they can act as a means of preliminary attack on incipient fires especially in regions where it is difficult to use heavy equipment.

The knapsack forest fire sprayers produced at present in the USSR and in other countries may be divided into three main groups according to their operating principle: manual, pneumatic and motorized. The most prevalent manual sprayer in the USSR is the knapsack forest sprayer, *RLO,* designed by G.A. Mokeev and A.V. E'ino. The most popular in other countries are the American sprayers, *Indiana, Hudson* and *Champion* with hard plastic, stainless steel, soft nylon, or neoprene reservoirs (Wajax Fire-Control Equipment, 1972). In the pneumatically operated sprayers, several modifications of the pneumatic knapsack sprayers of the *ORP* series (*ORP-A, ORP-B, ORP-V* and *ORP-G*), knapsack sprayers *ROOP* (*ROOP-4A, ROOP-4M*) and the chemical sprayer, *ORKh,* may be mentioned. Trials were conducted at the Leningrad Scientific Research Institute of Forestry on new types of sprayers, the high pressure in which is developed by means of Freon, and also on low-power internal combustion engine.

The knapsack forest sprayer *RLO* was first designed in 1936. It consists of two main parts: a shoulder bag and a hand pump with a double action spray gun connected by means of flexible tubing (Fig. 23). The bag contains 20 liters of fluid and is made of rubberized fabric resistant to the action of

Machines and devices for extinguishing fires with water

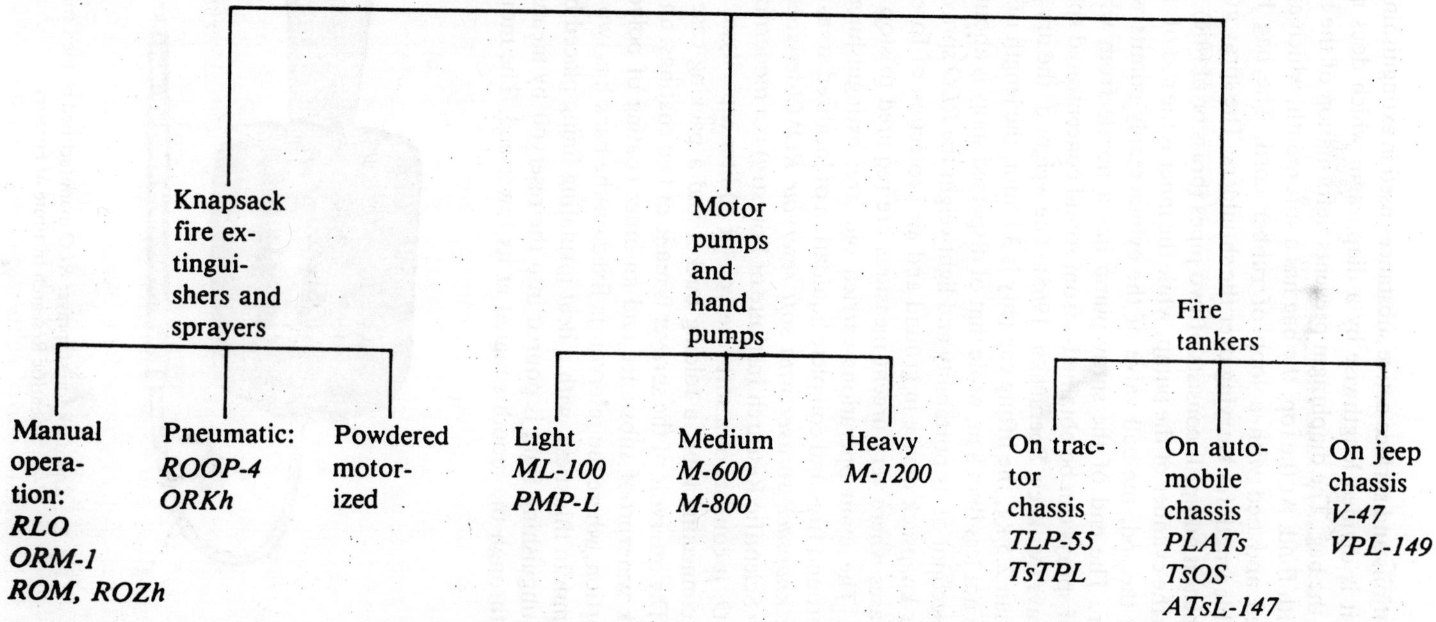

salt solutions and surface-active substances used in extinguishing forest fires. Inside, it is divided lengthwise by a diaphragm which does not reach the base of the bag. The diaphragm prevents deformation of the bag when it is filled with fluid. At the top, the bag has a soft mouth, which after filling up is folded and tied with a loop of rubber cord. The bag has straps of adjustable length for carrying over the shoulders. The spray gun is a double action piston pump. It consists of two pipes (brass or plastic), of which the outer is the cylinder of the pump, while the inner is the rod of the piston. On moving the rod, the ball valve of the cylinder only admits water into the cylinder. The end of the spray pump has a nozzle from which different kinds of sprays can be obtained—from round concentrated sprays to torch-like sprays (Mokeev, Speranskii, 1965). The weight of the dry sprayer is not more than 2.5 kg, the lifting capacity is 3 l/min, the length of concentrated spray is not less than 8 m, while that of dispersed spray is about 1.5 m.

On account of its portability and lightweight the *RLO* sprayer is the most popular knapsack device in ground and air protection of forests. It is used in all cases where very urgent measures are required to stop and suppress flames. The main operations carried out are: extinguishing the edge of forest ground fires and complete liquidation of localized areas of fire.

The *knapsack sprayer with soft reservoir ROM* (designed by the Far-Eastern Scientific Research Institute of Forestry) is a modernized version of the *RLO*. It consists of a soft reservoir of 20-liter capacity, a manual spray gun, a connecting tube, a folding bucket and a packing case (Bugai et al., 1972). The reservoir of the sprayer is made of two coatings: an outer coating of thick waterproof substance and an inner coating of polyethylene film. The portion, where the reservoir is fixed on the back has two layers of film of 8.4 mm in thickness, with a heat insulating lining placed between them. The extinguishing fluid is poured into the reservoir by means of a folding scoop through the mouth situated at its lower end. The mouth is covered

Fig. 23. Knapsack forest sprayer *RLO* manufactured by the Leningrad Scientific Research Institute of Forestry.

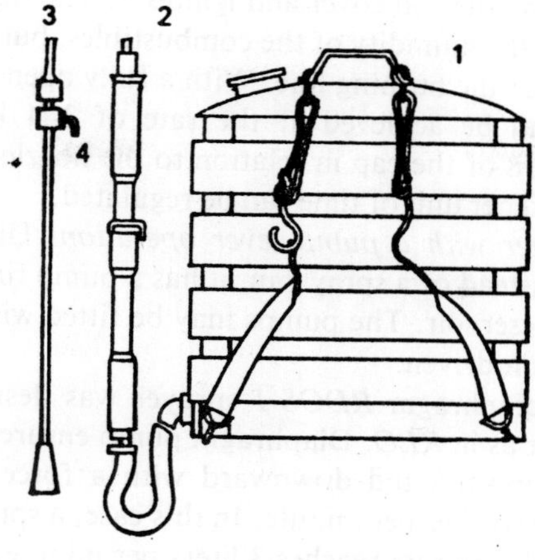

Fig. 24. Knapsack sprayer *ROZh* constructed by the Far-Eastern Scientific Research Institute of Forestry:
1—reservoir for fluid; *2*—manual spray gun; *3*—nozzle sprayer; *4*—safety valve.

with a lid fitted to a nozzle for connecting the tubing of the spray gun. During transportation of reservoir of the sprayer is wrapped around the spray gun and packed in the case. Dry weight of the sprayer is 1.6 kg. With a pressure of 2.5 atm in the spray gun and a nozzle outlet diameter of 2.5 mm, the fluid delivery rate reaches 0.12 l/s. A spraying device of such parameters ensures optimum dispersion of water droplets. The sprayer *ROM* is intended mainly as an equipment for fire parachutists and personnel of airborne landing units.

ROZh—knapsack sprayer with hard reservoir (designed by the Far-Eastern Scientific Research Institute of Forestry) is intended mainly for ground forest fire workers (Fig. 24). It can also be used as an ignition spray device. The reservoir of 18-liter capacity is made of stainless sheet steel. At the back, it has two projecting boxes, between which rubber straps are stretched, serving as shock absorbers during transport and protecting the back of the worker from contact with the cold reservoir. At the top, the reservoir has a mouth with a hermetically sealed lid of cylindrical form. One of the boxes contains the spray gun and the other contains the ignition device burner. The dry weight of the apparatus is 4 kg. The working principle of the spray gun is the same as the previous models of sprayers.

For using *ROZh* as an ignition device, the reservoir is filled with a fuel mixture (diesel oil and gasoline in the ratio 1 : 1), but instead of the nozzle of the spray gun, a burner is screwed on. The tube-shaped burner has a cock for regulating the flow rate of the fuel mixture and a fiberglass wick surrounded by a metal casing for protection of the flame from wind. The fuel mixture enters the burner through gravity and in the form of burning drops, flows down on the soil cover and ignites it. The flow rate of the fuel mixture depends on the humidity of the combustibles, but does not exceed 2 liters per kilometer of the burning line. With a fully open tap, the start of a ground backfire can be achieved at the rate of 3–4 km per hour. By changing the position of the cap in relation to the nozzle, atomization and flow rate of the fluid per unit of time can be regulated.

Knapsack sprayer with a pump lever operation. Unlike the sprayers described earlier, instead of a spray gun, it has a pump fixed directly on the framework of the reservoir. The pumps may be fitted with diaphragms, be gear-driven, or piston-driven.

The knapsack diaphragm *RDOS-1* sprayer was designed in 1938–39. The reservoir is soft as in *RLO*. Diaphragm pump ensures fluid blowout by swinging the rod upward and downward with a force of 6–8 kg and a frequency of 20–30 strokes per minute. In this case, a spray of 8 m length is formed, and the delivery rate reaches 3 liters per minute. For softening the ripple during the blowout, the pump has an air balloon. A trigger or some other type of control is mounted on the spray gun for regulating the flow rate of the fluid (Speranskii, 1955). Disadvantages of the *RDOS-1* sprayer are as follows: inconvenience in operation of the rod and its comparatively heavy weight (4.5 kg).

Knapsack sprayers with a manual trigger type pump did not find wide application due to the same disadvantages as were noted in *RDOS-1*.

One of the sprayers being produced at present is the *ORR* (*E'ra*) with a reverse action piston lever pump. It is meant for chemical treatment of gardens and vineyards, though it can be successfully employed for extinguishing ground fires of weak and medium intensity. This sprayer has a plastic reservoir which contains the piston pump and the air chamber. The shape of the reservoir is convenient for shoulder carrying. The gear of the pump is lever operated. With a force of 5 kg on the hand lever and with a wobble swing frequency of 30 per minute, the pressure created is about 3 kg/cm^2. The disadvantage of the apparatus is its comparatively heavy weight (4.5 kg) and the limited capacity of the reservoir (12 liters).

The *knapsack sprayer ORM-1* designed by *VNIIPO**) consists of a soft reservoir of 15-liters capacity, a piston pump with manual gear and a rod with spray gun. The filling of the fluid is carried out through a mouth in the upper part of the reservoir, while for switching on the suction pipe of the pump in the lower part of the reservoir, there is a special connection. Delivery of water from the reservoir is carried out by means of the piston pump. The piston has a multiple thread, which ensures rotary motion side by side with its reciprocatory motion. This lowers the resistance to its motion in the cylinder. The space between the walls of the cylinder and the

**VNIIPO*—Vsesoyuznyi nauchno-issledovatel'skii institut pozharnykh orudii (All-Union Scientific Research Institute of Fire-fighting Equipment)—Translator.

framework of the pump is used as the air chamber. Swinging of the handle of the pump does not require much effort. With a concentrated spray of seven meter length, the pump ensures a delivery rate of about 3 l/min. By regulating the nozzle of the spray gun, a torch-like spray of 1.5 m width can be obtained. In this case, the delivery rate of the pump is reduced to 2 l/min. The dry weight of the sprayer with the pump is 5.5 kg. The reservoir is provided with shoulder straps and a back cushion for convenience in operating the sprayer.

PNEUMATIC AND CHEMICAL SPRAYERS

The main disadvantage of manually operated sprayers is the necessity of expending large amounts of muscular energy, which, together with the high temperature of the fire, leads to rapid exhaustion. To eliminate this shortcoming, several designs of knapsack fire extinguishers and sprayers have been manufactured in which the blowout of the spray of the extinguishing fluid is effected either by means of compressed air forced during filling of the reservoir (manually or mechanically), or by means of gases formed as a result of chemical reaction in the reservoir itself.

Pneumatic knapsack sprayer ORP is produced in several models, *ORP-A, ORP-B, ORP-V* and *ORP-G,* which are distinguished from one another by a more improved design of the spray gun rod. The fluid reservoir is made of stainless steel. The framework contains a piston pump, the handle of which comes out through the lid of the mouth. By means of the pump, pressure up to 8 atm can be created inside the reservoir, which ensures a blowout of spray to a distance of up to 12 m. Preliminary hand-pumping of air at the site slightly facilitates operation of the sprayer, though it hardly solves the problem rendering the process automatic. This kind of sprayer, therefore, does not find wide application in forestry.

In order to mechanically create high-pressure in the cylinders (for example, by a motor pump, air pump used to inflate automobile tires, etc.), several modifications of pneumatically operated knapsack fire extinguishers and sprayers, *ROOP-4A, ROOP-4M,* were constructed at the Leningrad Scientific Research Institute of Forestry at the end of the sixties and beginning of the seventies (Speranskii, 1964). Mechanization of the laborious process of filling considerably facilitated the operation of the sprayer. Therefore, in recent years it has found wide application and is produced mainly for equipping stations for fighting fires with the help of chemicals.

Knapsack penumatic fire extinguisher and sprayer ROOP-4A (Fig. 25) consists of a shoulder-borne metal reservoir and a manual spray gun with a nozzle sprayer connected by means of a flexible tube. The dry weight of the apparatus is 7.3 kg, including that of the spray gun which is 1.1 kg. The reservoir of 20-liters capacity is made of stainless steel. It is made of two vertically placed cylinders connected with two pipes, of which the upper one

Fig. 25. Knapsack sprayer *ROOP-4* designed by the Leningrad Scientific Research Institute of Forestry.

serves as a handle for convenience of carrying on the back. One of the cylinders has a mouth for pouring the fluid and the lid is hermetically closed with a nozzle cap nut. At the back of the cylinder, there are straps of adjustable length and a soft-back cushion for the distribution of weight and protection of the worker's back from the cold cylinder holding the fluid.

The spray gun is a portable double action piston pump. Structurally, it differs only slightly from the spray gun of *RLO*. The nozzle-sprayer has two special mushroom-shaped nozzles with caps screwed on them. For ensuring a good spray, the outlets of the nozzle are situated at an angle to the walls of the guide cap.

The device can be operated in two ways: pneumatically, when the fluid is blown out through the spray gun under pressure of the air in the reservoir, and, manually, when it is blown out as a result of the operation of the spray gun. In the first case, an air pressure of nearly 2 atm is established in the cylinders and then by means of the charging unit (for example, motor pump *ML-100*) the fluid is compressed at a pressure of 6 atm. Subsequent filling is done only with water, while maintaining a residual pressure of not less than 1.5 atm in the cylinders. For delivery of the fluid to the fire edge, it is sufficient to open the tap of the spray gun and regulate the atomization of the spray.

In case of necessity, pressure can be created in the cylinders by means of the pumps used for inflating automobile tires. In this case, the fluid is filled first to 1/2 to 3/5 capacity of the reservoir which closes with the hermetically sealed lid. Then the connecting pipe is attached to the nozzle at the outlet of the engine of the vehicle and the pressure in the cylinder is raised to 6 atm. After disconnecting the nozzle and connecting the spray gun to the pipe, the sprayer is ready for operation. In the absence of compressed air, the reservoir of the sprayer is filled with fluid up to the top and spraying of the fluid on the fire edge is carried out manually by means of the spray gun.

Knapsack sprayer ORKh-3 (developed by the Central Air Force Base) structurally differs only slightly from the pneumatic sprayer, *ROOP-4*. However, the pressure in the cylinder is created by a chemical charge consisting of the following components: oxalic acid—100 g, potassium permanganate—34 g and bicarbonate salt—46 g. The sprayer is charged in the following sequence. First, the reservoir is filled with water to ¾ of its capacity. Then a chemical charge is placed at the bottom of the special cylinder which is already attached to the mouth of the reservoir, which is filled with 150 ml of water and the lid is screwed tightly, ensuring complete sealing of the cylinder. As a result of a chemical reaction accompanied by the liberation of a large quantity of carbon dioxide, pressure in the sprayer rises up to 5 atm in 2–3 minutes, ensuring ejection of the spray to a distance of 10 m. The residual pressure after all the fluid is consumed falls to 2 kg/cm^2. Chemical charges are packed in special bags which the worker carries with him. The total time required for filling the sprayer is five to eight minutes.

The advantage of the above method of creating pressure is its total self-sufficiency, i.e. independence from various types of charging sets and pumps. Its disadvantage lies in the lowering of the initial pressure developed in the cylinder by the charge if solutions of surface-active substances are used instead of water, and also the low accumulation coefficient of the reservoir in regard to the fire-extinguishing fluid.

While developing the idea of developing extra pressure in knapsack sprayers by a chemical method, V.G. Lorberbaum (1965) suggested the use of Freon 12 (dichlorodifluoromethane) for this purpose. Freon is packed under factory conditions in aluminum cylinders of 200 g capacity. The upper part of the cylinder has a check valve for filling and discharge. For using these cylinders in sprayers of the *ROOP-4* type, a holder in the form of a spherical brass strip and a screw for opening the check valve are fixed on the lid of the reservoir of the sprayer. The lid, along with the cylinder, is placed at the mouth and fixed with a nozzle cap nut. After opening the check valve, Freon is transformed from a liquid to a vaporous state, until the pressure in the reservoir corresponds to the given temperature. Thus, a pressure of 7.4 atm corresponds to a temperature of 20°C. In proportion to the release of the fluid in the cylinder, the free volume at once fills up with Freon vapors, while the pressure remains constant. This occurs until all the Freon is transformed into a vaporous state. With such a method of fluid ejection, the cylinder of the sprayer can be filled almost up to the top. The period of charging the apparatus does not exceed one-two minutes, while the cost of each charged cylinder is 30–35 kopeks.

MOTORIZED SPRAYERS

It is only in recent times that models of sprayers of this type have appeared in the country. The knapsack motorized sprayer (designed by the Leningrad Scientific Research Institute of Forestry) is constructed on the principle of the motorized agricultural sprayer. The main components of the sprayer are: a reservoir with a supporting frame and back straps, a small internal combustion engine, a detachable portable hydraulic centrifugal pump, suction and pressure hoses and light spray gun with a nozzle. The reservoir of 10-liters capacity is made of polyethylene. The pump develops a pressure of 5 atm, ensuring a fluid flow rate of 3.5–4 l/min and a range of concentrated spray up to 10 m. A detachable nozzle provides for delivery of an atomized spray of the extinguishing fluid. The dry weight of the sprayer is 10.1 kg, and the weight in working order is 20.7 kg.

The motor sprayer considerably facilitates the task of a fire worker, since it does not require expenditure of muscular energy for fluid ejection or complex accessories for filling it. At the same time, like any motorized device, it is more complex in operation. The gasoline engine on the back involves a greater fire risk, although it has a fire-extinguishing grid. One filling of the tank with fuel ensures continuous operation of the sprayer for 30 minutes. Considering the high efficiency of the sprayer and the limited capacity of the reservoir, it is recommended that it be operated in groups of two workers: one of them operates the device, while the other fetches water or other fire-extinguishing fluids, filling it into the reservoir. The latter operation can be carried out with the engine in operation. With such a pattern of labor organization, the results in extinguishing a fire edge can be improved three-to-four times.

MOTOR PUMPS

In the practice of extinguishing forest fires, motor pumps have found wide application. They differ in performance, fluid-head hose and weight characteristics. Depending on the pressure developed, motor pumps are divided into high-pressure (above 10 atm) and low-pressure (below 10 atm) pumps. According to their weight, they are divided into light (up to 30 kg) medium (from 30 to 100 kg) and heavy (above 100 kg). The equipment of the forestry department and the fire-chemical stations includes mainly light (*ML-100, PMPL*) and medium (*M-600* and *M-800*) motor pumps.

Motor pump ML-100 (designed by the Leningrad Scientific Research Institute of Forestry) is designed for extinguishing forest fires with water from natural and artificial reservoirs and for filling knapsack pneumatic fire extinguishers and sprayers with water or chemical solutions. It consists of either an engine or a gasoline-driven saw *Druzhba* along with air cooling, reduction gear and gear pump (Fig. 26). The pump has suction and flow pipes and two delivery hoses for filling fire extinguishers and sprayers of the *ROOP* type. It can develop pressure up to 10 atm, which enables utilization of the motor pump in rugged terrain. The delivery rate of the motor pump is 100 l/min for free discharge. It is advisable to use the motor pump *ML-100* in conjunction with pressure hoses of 26 mm diameter which permit maneuverability in forests (Speranskii, 1970). When the terrain is level, these hoses can deliver water up to a distance of 1 km, with the length of the jet at the lino end not less than 15 m. The light fire barrel has a charging unit at the end by means of which it is possible to quickly charge the knapsack fire extinguisher, *ROOP,* under a pressure of 6 atm directly near the fire edge situated far away from the water reservoir. The weight of the motor pump is 22 kg. It is marketed in a special container which can be conveniently carried by one person.

The motor pump PMPL (developed by *VNIILM**) is specifically designed for filling water in the tanks of fire engines. It has a high delivery rate (350 l/min) and can be used for delivering water by hoses under level terrain conditions, since the pressure developed (up to 4 atm) does not provide adequate pressure head in hoses. The engine of the motor pump is the gasoline-mechanized saw *Druzhba.* A distinctive structural feature of this motor pump, as compared to *ML-100,* is the low-pressure centrifugal pump (with a high delivery rate) and a propeller pump on a flexible shaft in the water intake hose, when setting up the pump above water level (in motor pump *ML-100,* this operation is carried out by the gear pump itself). In order to avoid a pressure loss in the hose, it is advisable to use this motor pump with hoses of 51 mm diameter. The weight of the motor pump (empty hoses) is 15 kg. This makes it possible to transport it to the remotest fire areas.

Motor pump M-600 is a device consisting of a two-stroke single-cylinder (12 hp) internal combustion engine and a centrifugal single-stage pump paired with a common shaft in a single frame with carrying handles. The water-cooled engine is driven by means of a pedal-type mechanism. For the initial intake from the water reservoir situated below the level of the motor pump, there is a slide-valve vacuum apparatus. The maximum suction height should not exceed 5 m. With the pressure of 5.5 atm, the pump ensures a delivery rate of nearly 520 l/min. The maximum pressure developed by the pump reaches 8 atm.

**VNIILM*—Vsesoyuznyi nauchno-issledovatel'-skii institut lesovodstva i mekhanizatsii lesnogo khozyaistva (All-Union Scientific Research Institute of Silviculture and Forestry Mechanization)—Translator.

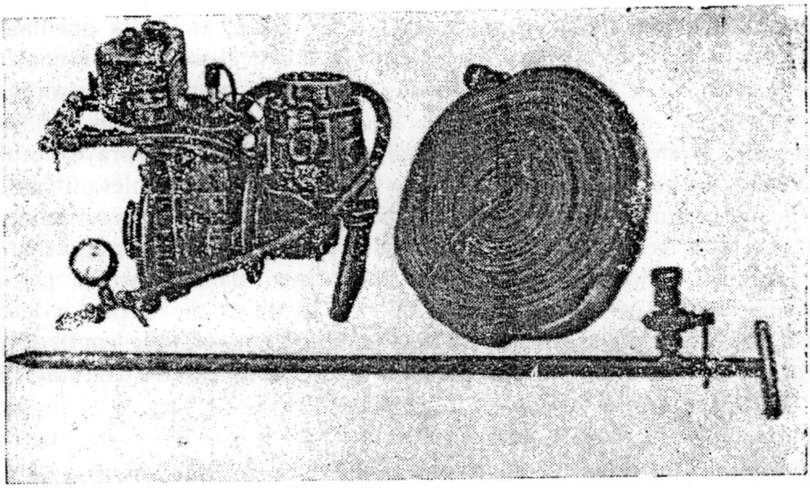

Fig. 26. Forest motor pump *ML-100* developed by the Leningrad Scientific Research Institute of Forestry with pressure hose of 26 mm diameter and long metal nozzle for extinguishing peat fires.

Usually, the *M-600* motor pump is used for delivery of water from reservoirs to the edge of large fires along main hoses of 66 mm diameter with a subsequent connection through a T-junction of narrower hoses (of 51 mm diameter). The weight of the motor pump along with the set of hoses of nearly 1 km length is about 500 kg. Therefore, this type of pumping apparatus along with the personnel should only be transported to the fire area by mechanized transport.

Motor pump MP-800 is mainly used by the fire brigades of cities, villages and settlements. However, it can also be used for extinguishing forest fires, especially in young woods, where it is necessary to deliver water with powerful long-range sprays. The 20 hp engine is gasoline operated, has two cylinders, with forced water cooling. The pump is centrifugal, single-stage, and panel-operated with a gas-jet vacuum apparatus for initial suction of water. The maximum suction height is 6 m. The engine and the pump are fixed on one frame, convenient for short distance transport. At maximum engine speed the pump develops pressure up to 8 atm. Such pressure permits the use of the motor pump *MP-800* for long distance delivery of water in rugged terrain. The net weight of the motor pump is 70 kg. Diameters of recommended pressure hoses are 66 and 51 mm. Delivery rate of the motor pump is 800 l/min.

Motor pump M-1200 is mounted on a single-axle trailer which can be transported by truck. The high capacity motor pump ensures delivery of a large quantity of water at a distance of over 1 km. The engine and the pump are covered with a bonnet with suction hoses piled on its sides. The bonnet has special flaps providing access to the control panel of the motor and the pump. All the disadvantages of heavy automatic pumps in regard to their use under forest conditions in the absence of roads are inherent in this type of pump. Motor pump *M-1200* and its modifications can be used in forestry only in exceptional cases, when delivery of water to the fire area is required in very large quantities (for example, when extinguishing fires which have penetrated into the forest floor and the peat horizon.

In the USSR, the first two types of motor pumps, viz. *ML-100* and *PMPL,* have found widest application in forest fire protection. Both, in the USA and in Canada, light motor pumps are preferred. However, in these countries emphasis is laid on obtaining high-pressure (nearly 20 atm) in the pressure hose. This is due to the necessity of using the motor pump in rugged terrain with the aim of compensating for pressure loss, which is inevitable when using hoses of small diameter.

The Canadian firm, *Wajax Equipment Ltd.,* has been manufacturing a light motor pump of the *Mark-3* type weighing 22 kg with an internal combustion engine of 9 hp capacity. The motor pump is characterized by a specific treatment of all units and components and its resistance to corrosion. The principal materials used in manufacture are aluminum alloy and bronze. The delivery rate of the motor pump, *Mark-3* under various pressures is:

Pressure, kg/cm^2	3.5	7.0	10.0	14.0	17.5	19.0	22.0
Delivery rate, l/min	333	288	229	162	85.5	49	Switched off

The motor pump of the *Mark-3* model is most commonly used and is often included in the complete set of self-propelled (power-fed) fire equipment (Wajax Fire-Control Equipment, 1972).

The American engineering firm *Waterous,* has been producing small multipurpose floating motor pumps. These motor pumps are installed on a plastic floating and work reliably without any intake hose in the water reservoir. The internal combustion engine develops up to 8 hp, the total weight of the motor pump and the float being 21.8 kg. The maximum spray range is 25 m, while the delivery rate is 568 l/min. Apart from its main purpose, viz. extinguishing forest fires, this type of motor pump is used for irrigation, washing cars, equipment, etc.

The *Pacific Marine Supply Company* has produced an experimental unit of power-fed fire fighting motor pumps, mounted on a light two-wheeled drive vehicle of the *Trail Breaker* make with a high cross-country performance. This vehicle has a special clutch, which enables each wheel to reach optimum rotational velocity independently of the other on encountering various obstacles. The vehicle easily operates in boggy areas. Standard equipment of the motor pump includes an instantaneous supply line for fuel and a portable gasoline tank of 19-liters capacity with a stop valve and handle for carriage by one person. Delivery rate of the centrifugal pump is 347 l/min under a pressure of 21.8 kg/cm^2. The total weight of the machine is 107 kg including that of the pumping device, which is 25 kg.

In the USA and in Canada, the tractor unit, Fire-Control Tank, (Model *TM-600*) and the mobile unit, *Fiberglass Slip-on Tank,* (Model *TFG-200*) (Wajax Fire-Control Equipment, 1972) have found a wide application. A high-pressure motor pump with a rubber hose on a self-winding reel is installed on the fire-extinguishing fluid tank. In the latter model, the tank of 800-liters capacity is made of anticorrosive material and has special semi-transparent windows for observing the water level. This unit is installed on a light vehicle on the 'pick-up' type and serves as a mobile means of suppressing incipient forest fires by a single person. In case of necessity, the body of the 'pick-up' can accommodate four fire brigade workers with manual implements (Wajax Fire-Control Equipment, 1972).

The most prevalent types of fire hoses in the American and Canadian Forest Conservation Services are flax-fiber hoses or hoses made of synthetic fiber with a latex coating on the inside. The diameter of the hoses is 38.1 mm. The Canadian firm, *Wajax,* which specializes in the production of forest fire equipment, produces flax-fiber hoses with a special impregnation against mold and decay as well as rubber coated hoses with a self-wetting quality. This property of the hoses protects them from damage by cinders. The hoses are designed for an effective pressure of 21 kg/cm^2 with loads up to 42 kg/cm^2. With a 30 m standard length of hose, its weight does not exceed 6 kg.

Hoses of the *Ambassador* make (USA) are of polyester with a double coating. They are much lighter as compared to single cotton, rubber-impregnated hoses and can withstand pressures up to 42 kg/cm^2. These hoses are resistant to mold and chemicals, but can be easily damaged by cinders (*Water Handling Equipment Guide,* 1969).

In forest fire control in the USSR, the most commonly used hoses are of flax-fiber and cotton with diameters of 26, 51 and 66 mm. In recent times, rubber-coated hoses are being used, although they are almost twice as heavy as flax-fiber hoses, which under forest conditions, renders them difficult to use.

As a rule, main lines from the water reservoir to the fire area are laid with large diameter hoses (for example, 66 mm). For laying operating lines branching from the main line, it is advisable to use hoses of a smaller diameter, ensuring better maneuverability near the fire edge. When using a highpressure motor pump, *ML-100,* the whole line is laid with hoses of 26 mm diameter.

Nonrubberized hoses are divided according to the working head (on which they are designed) into light (4–5 atm), normal (up to 12 atm) and reinforced (up to 15 atm). In order to be able to distinguish one type of hose from another, they are marked by longitudinal warps, i.e. stripes of colored threads. Light hoses do not have stripes, normal hoses have one stripe while reinforced hoses have two stripes at a distance of 10 mm from one another. Reinforced hoses are connected directly to the pump where maximum pressure develops. As the operator advances toward the fire, the pressure within the hose drops, therefore, standard or light hoses may be used. Non-observance of these conditions may result in the hoses being torn, if the pump is regulated at full capacity.

In the process of natural aging, and due to wear and tear hoses become less durable. In this respect, they are divided into three categories: I—hoses which are in use for less than 3 years and under pressure for less than 50 hours; II—hoses in use for 3 to 6 years and under pressure for 50–100 hours; III—hoses with a running period of over 6 years and under pressure for 100 to 150 hours. When transferring hoses from one category to another, the permissible effective pressure is lowered for light hoses by 1 atm, for normal hoses by 2.5 atm, and for reinforced hoses by 3 atm.

In recent years, a new type of light fire hose has been developed, for meeting forestry requirements. The hose consists of woven fiberglass sheathing impregnated with an elastic plastic coating which not only makes it resistant to decay, but also does not require drying. The smooth internal surface lowers pressure losses due to fluid friction against the walls of the hose, to the minimum. A hose with a diameter of 26 mm and a length of 20 m weighs about 3 kg and withstands pressure above 25 kg/cm^2.

In the USSR, *Rota* hose couplings, which ensure almost instantaneous connection, are used for joining hoses. In Canada and the USA, special threaded couplings are used for this purpose. More time is spent on connecting hoses by this method, but the joints are streamlined and are not affected by dragging of the hoses against various obstacles (for example, bushes).

A special connection, *OT-15* (*vodyanoi par*) is used in Canada for the transfer of water from the main hose into knapsack sprayers. It is made of a light alloy and has a one-and-a-half inch standard thread, which can be connected with hoses of the same diameter. A half-inch flexible pipe 0.75 m in length, pinchcock clamps, a nozzle and a rubber gasket are attached to the connection.

The service life of the hoses depends on their proper use. After use, the hoses should be cleaned of dirt, washed and dried. In Canada, a special device, C-2729, is used for washing hoses. It consists of two pipes of different diameters, placed one inside the other and screwed up at the ends by end-caps. While being washed, water from the pump flows through the flange of the outer pipe under pressure along the inner pipe. The hose is quickly dragged through the inner cylinder, where it is washed clean by thin spray of water from all sides. This device is made of a light anticorrosive alloy and weighs 4.5 kg. For convenience in operation, it is screwed with fastening brackets on to any support. For quickly stopping water at any place in the pressure hose, the hand clamp, B-2706, is used. The clamp is required when changing connections, nozzle and branches, and when extending hose pipes of 1–1.5 inch diameter. The length of the clamp is 28 cm and, its weight nearly 650 kg (Wajax, 1972).

For branching the main hose line near the fire edge, the joints with stopcocks are provided for simultaneous use by two workers. At the end of the operating lines, fire monitors are used for forming compact sprays. The spray gun (letter *B*) has a plug screwed on at the front end. A mushroom-shaped piece is screwed into the plug by means of a cross piece. The spray is regulated by the regulator, which is also screwed on to the plug. By turning the regulator, the spray can be gradually changed from a compact to a thin spray in the form of a cone with an angle of nearly 170°. Compact sprays are used for extinguishing fires in old stubs, deadwood, anthills and in hollows. Wet belts are created with a powerful compact spray ahead of ground fire fronts. For extinguishing fire at the edge, atomized sprays of water are usually employed.

FIRE TANKERS

The motor pumps described here ensure delivery of large quantities of water to fire areas. Working with heavy pressure hoses, filled with water, under forest conditions is sometimes difficult. The knapsack sprayer is a portable means of controlling ground fires but carries a limited supply of fire extinguishing fluid.

Fire tankers mounted on automobiles, tractors and cross-country vehicles (jeeps) have the combined advantage of high maneuverability and a large supply of fluid. Besides, they have a complete set of light forest fire equipment and, in individual cases, mounted soil working implements for constructing protective and supporting belts.

Forest fire tankers produced in the country include automobile-mounted *TsOS* and *ATsL-147, VPL-149* based on cross-country vehicles, and *TsTLP* and *TLP-55* based on tractors.

The *TsOS—tanker and detachable equipment* (developed by the Leningrad Scientific Research Institute of Forestry) is designed for extinguishing forest fires in regions with a developed network of roads suitable for motor transport (Speranskii, Shtuchkov, 1967). This set of fire-fighting equipment includes one or two (different hp) motor pumps with a set of hoses, a tank and portable equipment (Fig. 27). It is designed for transport on any automobile with a carrying capacity of not less than 2 tons. The tank is made of sheet iron of 2 mm thickness. The tubular framework serves as a rigid support for the tank. The tank is mounted on 8 castors, permitting easy rolling from the platform to the carriage of the automobile. On the top, the tank has 3 hermetically sealed hatches with rubber gaskets, and a

Fig. 27. Forest fire tanker *TsOS* with the complete set of equipment developed by the Leningrad Scientific Research Institute of Forestry.

mouth to fill it with fluid. The front and middle covers have sockets for fixing four knapsack fire extinguishers of the *ROOP* type and the ignition apparatus—*ZA-1M*. The front end of the tank has a specially designed bracket welded on it for fixing a crosscut saw and axes. The rear wall has stopper mounted on it for controlling the overflow of fluid and connecting the suction hose of the motor pump, *ML-100*. Seats for eight fire brigade personnel are fixed on the top of the tank.

The motor pump, *ML-100,* included in the set is provided with a device for the intake of a concentrated solution of surface-active substances-plus-wetting agents and its delivery into the pressure hose, through which the main stream of water flows. Filling the *ROOP* is carried out through delivery hoses, the motor being supplied with 26 mm fire hoses, 400 mm in length, packed in a special container. It is also provided with two barrels; one for forming concentrated and dispersed spray and the other for the supply of water, to peat layers.

Automobile fire tanker ATsL-147 is mounted on a high cross-country performance vehicle, *GAZ-66*. It is designed for quick transport of fire-extinguishing fluid (water with wetting agents), fire-fighting aids, and men of the fire brigade to the fire area. By means of this tanker, it is possible to extinguish the edge of forest fires with water or solutions of fire-extinguishing chemicals, construct protective or supporting belts and localize forest fires with exposed soil furrows, which help in final liquidation.

The forest fire tanker, *ATsL-147* (Fig. 28) has a tank for 1000 liters of water, a fire pumping unit, PNA 3/80, in a block with reduction gear, a semicovered metal cabin with seats for four members of the fire brigade, a

Fig. 28. Forest fire tanker *ATsL-147* mounted on a cross-country high performance vehicle.

carriage for portable fire-extinguishing aids and a mounted soil working disk implement with a remote controlled hydraulic system. The soil-working disk implement is mounted at the rear of the tanker. The disks cut into and turn over the top soil to a width and depth regulated by the road wheel. The fire pump of the tanker operates from a power takeoff shaft on the vehicle. The pump is controlled from a cabin, where control levers are situated. Intake of water from the reservoir is effected by means of a vacuum system, while the supply of water from the tank is through a smooth rubber hose of 60 m length by means of a pump. For delivery of water at long distances, there is the motor pump, *PMPL,* with a set of hoses 51 mm in diameter. Besides, there are 4 sprayers of the *PLO* type for extinguishing the rear and flank edges of ground fires and for complete liquidation of the remaining fire sources as well as peat barrel, *TS-1* for extinguishing peat fires.

The forest fire tanker, *ATsL-147,* is designed as an equipment for forest fire chemical stations situated in conservation zones with a comparatively dense network of unsurfaced roads for vehicular transport.

The *cross-country forest fire tanker VPL-149* (designed by *OKB** fire engines) is intended for quick transport of fire brigade workers, firefighting aids and fire-extinguishing fluid to forest fires in areas where there is an absence of roads or to places where vehicular transport is difficult. It has a protected body lined with aluminum sheets, with ventilation through the hatches in the roof and front panel. The rear door has two flaps with

**OKB*—Opytno-konstruktornoe byuro (Experimental Design Office)—Translator.

Fig. 29. Cross-country forest fire tanker *VPL-149.*

locks and folding steps. Inside the body, two tanks with a total capacity of 1,000 liters are installed, with four semi-upholstered seats for the fire brigade workers mounted on these tanks. The complete set of fire equipment of the cross-country tanker includes a motor pump, *PMP-L,* 4 knapsack sprayers *RLO,* a motor pump *Druzhba,* an ignition apparatus—*ZA-FK,* an elastic tank of 1,000-liters capacity, a thermos for drinking water, spades, axes, a double-handled saw, first-aid kit, etc. The vehicle is equipped with a shortwave radio set with an operational range of not less than 50 km on level terrain. The vehicle can develop a speed of up to 50 km/h on level terrain and can surmount inclines and gradients of up to 35°. Besides, without the fire-extinguishing fluid on board, the vehicle can cross water channels (by floating) and boggy regions of and can also fell individual trees up to a diameter of 15 cm, which obstruct its movement. The vehicle, *VPL-149*, is equipped with the same soil-working implement as the tanker, *ATsL-147*. The soil-working implements in both the vehicles are interchangeable, and the hydraulic change is standardized at 70% of the total load (Fig. 29).

The *tractor-mounted forest fire tanker, TsPLT-2* (developed by the Far-Eastern Scientific Research Institute of Forestry) is a unit of detachable anti-fire equipment mounted on a caterpillar tractor *TDT-60* or *TDT-75*. It includes: two tanks of fire-extinguishing fluid each of 1500-liters capacity, a gear pump, *NShN-600,* with a delivery rate of 600 l/min, a set of suction pressure hoses and a fire barrel, a motor pump, *ML-100,* a gasoline-motor saw—*Druzhba,* an ignition apparatus, 6 knapsack sprayers (2 *PLOs* and 4 *ROOPs*), shovels, axes, buckets and other equipment. A chain ripper-cum-dragharrow is secured at the back of the tractor. The complete kit of the tanker is only mounted on tractors of the above makes during fire risk periods (Fig. 30). The tanker is designed for controlling forest fires within a

Fig. 30. Tractor-mounted forest fire tanker, *TsPLT-2,* developed by the Far-Eastern Scientific Research Institute of Forestry.

radius of 25–30 km from the base, under rugged conditions of the Far-Eastern Region of the Soviet Union. It is designed for transporting water and other fire-fighting aids to the forest fire area, extinguishing fires with water from fire barrels and knapsack sprayers, laying barriers and supporting belts with fire-extinguishing fluid and exposed soil strips by means of the ripper-cum-drag-harrow. The equipment is provided with 6 seats for transporting fire brigade personnel.

The *forest fire unit TLP-55* (by the Leningrad Scientific Research Institute of Forestry) is mounted on a caterpillar tractor, *TDT-55,* and is designed for controlling forest fires in logging areas (Fig. 31). A complete set of forest fire equipment of the unit includes: two tanks with four seats for fire-brigade workers, a pumping device with a gas spray vacuum apparatus, a plowlike soil-working implement, a bulldozer and portable forest fire equipment (motor pump *ML-100* with a set of hoses, ignition apparatus, gasoline motor saw—*Druzhba,* 4–*ROOPs* and 4-*RLOs,* shoulder containers for transporting fire hoses, peat barrel *TS-1,* shovels, axes, etc.). Capacity of the tank is 1,000 liters. The soil-working implement has V-shaped blades of the plow-trencher *PKLN-500* type. Lifting and lowering of the implement is effected by remote control from the driver's cabin. In dry sandy and sandy loam soils, it is possible to create protective exposed soil belts of 2 m width at the rate of 1.4 to 2.2 km/h by means of this implement, remove obstacles in the direction of movement of the unit, provide access to forest reservoirs for water intake and, in some cases, to set up wide protective belts. Besides, the blade of the bulldozer serves as a

Fig. 31. Forest fire unit *TLP-55* developed by the Leningrad Scientific Research Institute of Forestry.

balance weight for counterbalancing the plowlike implements.

Fire tankers on wheels and with tracks are widely used in forest fire protection services in Canada and the USA and also in several European countries (France, Federal Republic of Germany, Poland). The equipment of forest fire stations in the USA and Canada includes mobile tankers of 1,100 to 3,000-liters capacity. The latter are installed on 3-ton vehicles with a pump operating from the power takeoff shaft. The pump is used for filling the tank as well as for delivering water to the fire area along a hose line. The tanker has nearly 360 m of fire hose of 38 mm diameter, a canvas bag of 185-liters capacity for intermediate fluid transfer if required, and a complete set of manual fire-fighting implements (knapsack sprayers, ignition apparatus, shovels, axes, motor pump, etc.). Some private logging and timber sorting firms have tankers with a capacity of nearly 9.5 thousand liters, which can be quickly mounted on heavier timber transport vehicles. Along with the tank, a duplex pump is installed on the vehicle and a reel with 600 m of hose of 38 mm diameter is wound around it.

The tracked forest fire unit *Bombardier* has high tactical features and specifications (Nikolayuk et al., 1967). On a caterpillar tractor of the same make, a tank of 2 m^3 capacity is fitted with a self-contained pumping device, which develops a pressure of up to 13 atm. Sometimes such a unit is equipped with a small plow for constructing protective belts. The capacity of the engine is 85 hp, which develops a speed on unsurfaced forest roads of up to 40 km/h and easily surmounts steep slopes and other obstacles. Broad rubber-metal tracks do not damage the surface root system of trees and ensure good performance of the unit on rough roads. This *Bombardier* unit is widely used for controlling forest fires in the central and eastern provinces of Canada.

In Australia, the firm *Softwood Holdings,* has developed a fire tanker for emergency fire fighting in man made plantations. A *Bedford* truck with a four-wheel drive for high cross-country performance is used as the basic model. A powerful pump with a delivery rate of 1134 l/min fills a tank of 7,000-liters capacity in 6 minutes. A monitor with a fire hose and a large sized fire barrel is installed in the body behind the driver's cabin. On each side of the body, there is a bracket for the monitor, which is used when required for extinguishing fires in tall trees, hay stacks, burning structures, etc. For extinguishing small fire areas, two revolving reels with rubber hoses of 19 mm diameter and 30 m length are provided. These are equipped with pistol-type barrels with apertures of 6.3 and 7.9 mm. These hoses are intended for extinguishing fires directly from the vehicles. Besides, two short hoses of 2.4 m length each are provided with the universal fire barrels which deliver water in the form of a compact spray, splash or water mist. The mist is created around and above the mobile tanker for protecting it from fire in emergency situations. The sprinkler hose installed in the front of the vehicle is intended for wetting a belt of 2.5 m width and is used, for example, in extinguishing grass fires (*Australian Forest Industries Journal,* 1973).

The high boards of the body protect fire workers from intense heat radiation. In addition to the water screen, a provision is made for sprinkling the long chain of edge fires and the fuel tank in order to avoid their overheating above the body. Six knapsack sprayers of the *Indiana* type, tarpaulin hoses and boxes with instruments are accommodated in special cupboards with an internal lighting system. There is a 12 volt lighting lamp on a flexible cord, signals on the rear part of the body for communicating with the driver, an electric fuel pump, an enclosure chain behind the body for ensuring safety, an intake hose of 6 m length, a water level indicator in the tank, a tap for filling knapsack sprayers, and a radio set.

The New Zealand firm, *Forest Products Kinleys,* has manufactured a mobile pumping device, mounted on a high-speed vehicle, with a delivery rate of 4,740 l/min. With effective pressure of 7 atm and a delivery rate of 3,800 l/min it can deliver water through 5 hoses. The vehicle is painted white for better identification in smoke, fog and artificial lighting. The maximum speed of the vehicle with a total weight of 1 t is nearly 200 km/h (*The New Zealand Timber Journal,* 1971).

Estimation of Number of Workers Equipped with Knapsack Apparatus for Extinguishing Forest Ground Fires

In the operation of controlling forest fires, it is essential to know what kind of forces and means should be directed toward extinguishing freshly detected fires. Overestimation of the rate of propagation of forest fires over an area leads to an excessive utilization of effort and means while underestimation leads to a situation when fire may get out of control due to loss of time.

A method developed by the Leningrad Scientific Research Institute of Forestry (Vonskii, 1971) helps in carrying out in two-or-three minutes, a rough estimate of the number of workers, equipped with knapsack apparatus required for extinguishing the edges of forest around fires in the initial period of their propagation (within 3 hours). The principle of calculation can also be used for planned determination of personnel strength and territorial distribution of fire brigades by taking into account the frequency of occurrence of fire over delineated areas. The basis of this method involves the knowledge of data on the rate of propagation of forest fires according to perimeter and area, depending on weather conditions and performance of workers equipped with *ROOP* type sprayers.

For practical calculations, forest stands of various types with similar pyrological characteristics are combined into groups:

I. Types of forest for which data on the propagation rate of ground fires is available:

lichen-covered pine forest

drained whortleberry pine forest (pure composition stands)

II. Types of forest to which these data may be extended:

lichen-moss pine forest

cranberry pine forest

drained whortleberry and whortleberry-sphagnum pine forest stands, (pure and mixed with spruce and hardwood trees)

drained whortleberry and whortleberry-sphagnum spruce forest stands, (mixed with pine and hardwood trees).

The rate of spread of the fire edge for various tactical elements (rear,

TABLE 17. LENGTH OF GROUND FIRE EDGE DEPENDING ON WIND VELOCITY, HUMIDITY OF COMBUSTIBLES AND DURATION OF COMBUSTION TAKEN FOR WEAK COMBUSTION INTENSITY

Wind velocity, m/s	Humidity of combustibles, %	Length of fire edge, m, depending on combustion duration, h					
		0.5	1.0	1.5	2.0	2.5	3.0
		Heather-covered pine forest					
0.0		60	120	180	240	300	360
0.5		80	160	240	320	400	480
1.0	10	150	300	450	600	750	900
1.5		220	440	660	880	1100	1320
2.0		400	800	1200	1600	2000	1240
0.0		45	90	135	180	225	270
0.5		55	110	165	220	275	330
1.0	20	110	220	330	440	550	660
1.5		155	310	465	620	775	930
2.0		280	560	840	11240	1400	1680
0.0		25	50	75	100	125	150
0.5		35	70	105	140	175	210
1.0	30	60	120	180	240	300	360
1.5		95	190	285	380	475	570
2.0		145	290	435	580	725	870
		Lichen-covered pine forest					
0.0		65	130	195	260	325	390
0.5		80	160	240	320	400	480
1.0	10	145	290	435	580	725	870
1.5		195	390	585	780	975	1170
2.0		445	890	1335	1780	2225	2670
0.0		45	90	135	180	225	270
0.5		60	120	180	140	300	360
1.0	20	105	210	315	420	525	630
1.5		135	270	405	540	675	810
2.0		200	400	600	800	1000	1200
0.0		30	60	90	120	150	180
0.5		35	70	105	140	175	210
1.0	30	50	100	150	200	250	300
1.5		70	140	210	280	350	420
2.0		120	240	360	480	600	720
		Drained whortleberry pine forest					
0.0		70	140	210	280	350	420
0.5		80	160	240	320	400	480
1.0	10	160	320	480	640	800	960
1.5		225	450	675	900	1125	1350
0.0		50	100	150	200	250	300
0.5		65	120	180	240	300	360
1.0	20	85	170	255	340	425	510
1.5		165	330	495	660	825	990
0.0		35	70	105	140	175	210
0.5		40	80	120	160	200	240
1.0	30	50	100	150	200	250	300
1.5		65	130	195	260	325	390

front and flanks) is different and, therefore, data on the length of the edge is given for weak combustion intensity (in the rear). Thus, the calculated fire edge will always be slightly longer than the actual edge (Table 17).

Determination of the humidity of combustibles (of soil cover) is quite complicated. However, investigations have established that a definite dependence exists between fire risk according to weather conditions, and the humidity of combustibles.

The degree of fire risk from weather may be expressed by the drought index (PZ), which is calculated for each day by the formula

$$\mathrm{DI}_n = (\mathrm{DI}_{n-1} + d_{n-1})\,K_n,$$

where DI_n is the draft index for the morning of the current day, $\mu\delta$, DI_{n-1} the draft index for the morning of the preceding day, $\mu\delta$, d_{n-1} the daily average moisture deficit of the preceding day, $\mu\delta$, and K_n the rainfall coefficient established for the current day.

Thus, the draft index is the same as the total daily average values of the atmoshpheric moisture deficit multiplied by the rainfall coefficient. The value of DI is in inverse proportion to the quantity of rainfall:

Quantity of rainfall, mm	Nil	0.1–0.9	1–2.9	3–5.3	6–14.9	15–19.9	≥20
Rainfall coefficient K_n	1	0.8	0.6	0.4	0.2	0.1	0

This relationship has been established by the Hydrometeorological Centre of the USSR according to data obtained by the Leningrad Scientific Research Institute of Forestry. The procedure for calculating draft index for each day of the fire risk season is explained in more detail in the manual *Metodicheskie Ukazaniya po Sostavleniyu i Primeneniyu Mestnykh Shkal*

TABLE 18. RATIO BETWEEN THE HUMIDITY OF THE MOSS-LICHEN STORY OF VARIOUS TYPES OF FORESTS AND THE VALUE OF THE DRAFT INDEX

Forest type	Dependence of draft index on humidity of lichen-moss story, %		
	10	20	30
Pine forests:			
heather-, lichen- and lichen-moss-covered	40	21–40	10–20
cranberry-covered	50	36–50	21–35
drained whortleberry	60	41–60	31–40
whortleberry-sphagnum	100	81–100	71–80
Spruce forests:			
drained whortleberry	90	51–90	41–50
whortleberry-sphagnum	100	81–100	71–80

Pozharnoi Opasnosti v lesu (Instructions on Methods of Composing and Applying Local Fire Risk Classification in Forests).

For determining the theoretical humidity of a moss-lichen story of a particular type of forest, it is sufficient to find the value of the draft index and to determine its value according to Table 18.

As mentioned earlier, the wind velocity in an open space and under a forest stand canopy is different and depends on many factors. For practical calculations, the wind velocity under a wooden canopy can be established according to data of the nearest meteorological station or according to the visual estimation scale of the wind force given in manuals for hydrometeorological stations and posts. Assumptions on determining the wind velocity under a forest canopy in relation to the wind velocity outside the forest limits are given below:

Wind velocity:					
outside forest limits	up to 1	1–3	3–4	5–6	>6
under the forest canopy in heather-, lichen-, lichen-moss- and cranberry-covered pine forests	0	0.5	1.0	5–6	1.0
outside the forest limits	up to 3	3–5	5–6	>6	—
in whortleberry- and whortleberry-sphagnum-covered pine and spruce forests	0	0.5	1.0	1.5	—

By using data of wind velocity under a forest stand canopy and humidity of combustibles, it is easy to find the theoretical length of a ground fire edge.

When extinguishing fires by means of the *ROOP*, the best performance can be achieved if the brigades operate in teams of two persons. One of them walks in front and beats the flames at the fire edge with a dispersed spray and the other puts out the remaining areas. The performance of the team depends on the intensity of combustion, which in its turn, will be different in different types of forests (Table 19).

Data in Table 19 has been determined chronometrically while extinguishing ground fires of varying intensities. As compared to actual data, it has been reduced by 25%, taking into account various unforeseen circumstances occurring during fire-fighting operations. The maximum time of continuous operation of the *ROOP* with one water charge is 5 minutes, while the capacity of a worker team with these sprayers is from 23

TABLE 19. LENGTH OF THE GROUND FIRE EDGE WHICH CAN BE SUPPRESSED BY A TEAM OF TWO WORKERS EQUIPPED WITH *ROOP*

Type of pine forest	Length of fire edge, m, with combustion intensity		
	severe	medium	weak
Heather-covered	80	100	130
Lichen-covered	60	100	160
Drained whortleberry	—	60	120

to 42 m/min (taking into account the 25% reduction in capacity). The time spent by the workers on one cycle, i.e. from one charge to the next (including the time of transfer from the charging point to the fire area, water charge itself and extinguishing) depends on the distance between the charging point and the fire area and is developed according to chronometric data:

Distance from charging point to fire area, m	100	200	300	400	500	600	700	800	900
Time spent on one cycle, min	15	20	25	30	35	40	45	50	55

Proceeding from this data, the number of Z teams required for extinguishing a forest ground fire with an expected length of the edge, L, can be determined by the formula:

$$Z = \frac{L + \pi l_1 t}{l(\pi + 1)}$$

where π is the number of recharges of the *ROOP* required for complete suppression of fire; l_1 the length of the fire edge, which can be tackled by the team in one operating cycle, in m (see Table 17), t the time spent by the team on one charge cycle of *ROOP* before recharging, in minutes. The value of π is the quotient of the expected length of the fire edge, L, by the length of the edge which can be controlled by the team of workers in one operating cycle, l. The number of teams to be sent to the fire area to liquidate it without a repeated recharge of the *ROOP* is determined in the same manner:

$$Z \geqslant \frac{L}{l}.$$

Thus, in order to carry out indicated calculations for a realistic operation, it is essential to determine: the fire area (by the method of intersection from the observation posts) and the assessed characteristics of the forest stand where it has occurred (from the plan of the forest stands); the time required for transporting workers with the *ROOP* to the fire area, taking into account the distance to the fire area and the speed of transport for carrying workers and fire-fighting equipment; the wind velocity (according to the data of the meteorological station or the visual estimation scale); and the value of the draft index. Using the above tables and formula, it is comparatively easy to establish the personnel strength and the means required for extinguishing the incipient fire. However, these can serve only as general guides, and should be determined, taking into account the local forest vegetation and weather conditions. When using other technical aids for suppressing fires in fire-fighting practice, these indexes will be different, although the basic methods of calculation remain the same.

Organization of Protective Belts and Trenches

The main function of protective belts and trenches is to serve as obstacles to the spread of ground or underground fires. They can also serve as a supporting boundary for the start of a ground backfire.

A protective belt (through exposed soil belt or canal) ensures separation of combustibles from a combustion zone or renders it noncombustible (through treatment of combustible materials with water or solutions of fire-extinguishing chemicals). Often, protective belts are set up in order to localize ground fires which have engulfed large areas of forest territory. Besides, they are essential for a more reliable localization of those fires, movement of the edges of which has been checked by beating, by covering with soil or by spraying with water.

Tractor plows, *PL-70, PLP-135,* bulldozers, soil-working implements of the fire tanker, *ATsL-147,* and the cross-country fire vehicle, *VPL-149,* horse plows, explosives, etc. are used for creating exposed soil protective belts. Use of soil-working elements on tractor or automobile traction results in marked improvement in the application of methods and ensures a saving of time.

Soil-working implements for creating protective belts should meet the following requirements: they should be easy to handle and be quickly connected with the tractor, during the operation of the device they should require minimum additional manual handling, they should have a minimum traction resistance and minimum possible deviation in a horizontal position at the widest angle from the tractor, they should be capable of operating in rugged terrain and should be simple in design and sturdy in construction.

The most reliable protective belt is made with the use of a pair of bulldozers. The pattern often used in the USSR and Canada for making such a belt is as follows: the first bulldozer fells large trees in the track of the intended belt and removes them to the side with the blade raised above the ground; the second bulldozer exposes the upper layer of combustibles up to the mineral layer; the latter operation can be carried out by two methods: either directly with an obliquely placed blade which ensures removal of the layer of combustibles to the side or in a herring bone manner, i.e. by running backward and forward at a certain angle to the axis of the belt. The first method should be used in light sandy soils with underdeveloped soil cover and the second method in heavily cluttered forest stands of the moss-covered type (with cranberry, whortleberry and haircap-moss covers). In both cases, the combustible is removed to a side away from the fire. Under average conditions, experienced tractor operators lay out, in one hour, up to 600 m of a protective belt of about 3 m width. The necessity of operating bulldozers in pairs arises not only due to a desire to improve their performance and save time (this is achieved by a division of operations), but also by the tactical situation likely to develop unexpectedly in the fire area. The first bulldozer, besides felling trees in its tracks, plays an exploratory role and, therefore, often gets stuck in damp depressions while the second bulldozer quickly pulls it out and saves time. This situation should also be taken into account when developing protective belts with other tractor units.

In underground fires, especially in forests with a developed peat horizon, protective trenches are dug. The trenching plows *PKLN-500, LKN-600, LKA-2A,* assembled with tractors of 3 t capacity and above, are used for this purpose. The maximum depth of the trenches constructed by these implements is up to 60 cm with a surface width of 1.5 m. When the depth of the peat horizon is more than 80 cm, the excavator *E-352* is used. The trench is dug up to the mineral soil or moist peat layer incapable of combustion. In the absence of digging machines, protective belts (more seldom, trenches) are created manually. In manual operation, the soil cover and forest floor are raked up to the mineral soil at the side away from the fire. Width of the belt is 60–80 cm. The working implements include: rakes, hoes, metal brooms, spade, etc.

If the fire edge is stopped, the protective belt is laid directly near it. With a moving fire edge, the protective belt is laid at some distance from it, taking into account the rate of advance of the fire.

One of the main parameters of the protective belt is its width. It can be from 0.5 (Antsyshkin, 1957) to 4 m and above (Kurbatskii, 1962). The width of the protective belt should be regarded as a function of combustion intensity and the angle of inclination of the flame to the horizontal plane. It should be ensured that on the approach of the flame, the possibility of ignition of the combustibles on the other side of the belt due to the flow of inflammable gases and radiation, is excluded, although sparks flying across the belt in such cases cannot be avoided.

With bulldozers and tractor plows operating in the fire area, the width of the protective exposed soil belt is determined by the operating width of the implement, and cannot be changed depending on the combustion intensity or upon the tactical elements of the fire edge. In order to increase the width of the barrier belt in one passage of the tractor, special belt-laying machines are used.

The milling-and-belt-laying device *PF-1* (developed by the Leningrad) Scientific Research Institute of Forestry) operates within the speed limits

Fig. 32. Milling-and-belt-laying implement *PF-1* developed by the Leningrad Scientific Research Institute of Forestry.

and with active control from the power takeoff shaft of the tractor. It is attached to tractors *LKhT-55, T-4* equipped with hinges and also with *T-74* and *DT-75* with reduction gears (Fig. 32). The working device of the implement consists of two transverse milling scarifiers—568 mm in diameter. In one passage of the tractor, the belt-layer constructs a furrow 120 cm in width and up to 20 cm in depth with soil ejection on both sides for 3–4 m. Thus, the total width of the belt comes to 7–10 m, with a forward working speed of the unit from 1.5 to 3.5 km/h. The milling depth can be regulated, taking into account the character of the soil and the depth of the forest floor. The belt-layer can successfully operate in different types of forest with sandy, sandy loam and light sandy loam soils within the operating limits of the tractor. Distribution of the ejected soil along the width of the belt is regulated by the casing wings. On approaching the cover which is sprinkled with soil, the flame suddenly subsides. If smoldering continues in the layer of the soil covered floor, it is stopped by the wide furrow (Valdaiskii, Kodyanov, Chukichev, 1972).

The efficiency of protective belts created by fire extinguishants depends not only in their width, but also on the degree of impregnation of the combustibles with chemicals. Besides, chemicals ensure quick stoppage of flame combustion and smoldering, exclude the possibility of fire revival, delay the spread of fire for a definite interval, dissolve well in water and are nontoxic and inexpensive.

Protective belts are laid in uncluttered areas with a minimum of vegetation. The normal recommended width of a belt is 1 m, whereas for densely covered areas with a wind velocity under the forest canopy of over 3 m/min, 2 m soil cover is advised. Components on the protective belt should be uniformly moistened with fluid. If a small area of the belt is missed out or insufficiently moistened, it can serve as a 'bridge' or gap for movement of the flame through the belt.

In cranberry- and whortleberry-covered forests, the required dose of the solution is not less than 1 liter per square meter; in lichen-moss and heather covered forests, it is 0.5–0.75 liter per square meter; in boggy sphagnum-covered forests with ledum and other bushy plants, it is 1.5 liters per square meter, while in dense areas of forest, with all types of forest cover, the dose is not less than 2 liters per square meter. Development of 100 m long and 1 m wide protective belt with a solution dose of 0.5 liter per square meter for a knapsack sprayer requires 15–18 minutes, while with a solution dose of 1 liter per square meter, the time required is 30–35 minutes.

The time lag for the belts to actuate depends on the properties of the fluids used for treatment of the combustibles in the belt. When organizing protective belts before ground fire with actuating time being 1 hour, water with the addition of wetting agent is an effective fluid in most inflammable types of forests. In case of a need for the creation of fire-retarding belts with an actuating time of up to 24 hours, a 20% solution of calcium chloride with the addition of a 0.5% wetting agent such as *OP-7,* should be used. With the addition of a wetting agent in a viscous solution of calcium chloride, the actuating time of the belt increases up to 48 hours and more. The same result can be obtained if the belt is treated with phosphorus pentoxide in a powdered form at the rate of 0.1 kg/m^2. A belt with aero-mechanical or chemical foam has high fire-retarding properties, though the 'life' period of the foam is only 15–20 minutes, after which the belt remains wet.

In Canada and the United States of America, sodium and calcium borates, as well as the powder known as *Firebrake,* are widely used for creating protective belts. When mixed with water in the proportion 1 : 2, it forms a thixotropic emulsion which in the stationary state remains viscous and, on stirring, becomes fluid again. The solution is heavy and penetrates well into the forest floor, forming a white film after drying, which retards flame combustion. The rate of solution per square meter of the belt is from 1.2 to 2.1 liters (Tamarkin, 1966). Aqueous suspension of bentonite clay in the ratio of 1 : 10 is sometimes used for this purpose. The rate of solution per square meter of the belt under treatment is 1 liter on an average.

In recent times, a thickener named 'Herald', in the concentration of 0.5–1.0 kg per 100 liters of water has been used in the USA and Canada as a fire-retarding agent. This is a sodium salt of alginic acid extracted from marine algae. On adding a small quantity of calcium chloride to viscous water, a jelly-like mass is obtained which adheres well to bushes and cover. According to American scientists, who have tried this thickener under laboratory conditions, it increases the efficiency of water by 4 times (Timberman, 1968).

In 1958–1962, the Forestry Commission of Great Britain conducted comparative trials of sodium-calcium borate, an aqueous mixture of bentonite clay, solutions of primary and secondary ammonium phosphate and a thickener (an aqueous solution of sodium salt of alginic acid). The trials were conducted during high fire risk periods in herbaceous, heather-covered forest stands and also in dense spruce plantations. As a result, the following widths were recommended for protective belts: flanks and rear of fire—0.9 m; lee fire before the front—1.8–2.7 m; supporting lines for annealing and for controlled combustion of slash in felling areas—0.5–0.9 m. All the solutions showed good results even in dry condition. The film of borates inhibited combustion for several weeks while the film of bentonite, ammonium phosphate and the thickener was effective for some hours. Ammonium phosphate is a good fertilizer while borates are total herbicides.

In heather-covered woods, viscous water was found to be the most effective. Belts of 0.9 m width treated with the thickener at the rate of 1.1 liters per square meter stopped the motion of flame at the edge while ammonium phosphate was ineffective with a belt width of 2.7 m and a rate of solution up to 2.2 liters per square meter (Tamarkin, 1966).

An increase in the efficiency of the protective belt by an increase in its width is technologically simple, but a more laborious process. At the same time, the width of the belt can be reduced to the minimum if a short barrier is placed before the fire edge, diverting incandescent gases upward and fulfilling the role of a screen preventing the heating of combustibles due to convection and flame radiation. Experiments conducted in 1966–67 by the author in collaboration with P.A. Gubin confirmed the high efficiency of the sheet barrier. An asbestos barrier strip 1 mm in thickness and 40 cm in height with the same height as of heather, effectively stopped a ground fire of high intensity with a wind velocity of 2.5 m/s and gusts up to 4 m/s. Under these conditions, the ratio of the height of the barrier to the width of the protective belt with the same efficiency reached 1 : 8.

Method of Ground Backfire

The start of a ground backfire is an ancient method for extinguishing forest fires. It consists mainly in burning out soil combustibles in an area, before the fire edge of a severe ground or crown fire reaches it. In spite of the development of new, original technical means of extinguishing fires, this method has so far not lost its significance. On the contrary, due to the possibility of mechanizing the process of constructing a supporting belt, it has developed further.

Usually, a backfire is started in a direction opposite to the frontal edge of the fire, less often against the flank edges and sometimes against the rear edge; although in this case, the flame travels with the wind. At the frontal and flank edges of a ground fire, combustion is stopped due to the absence of combustibles in its course which disappear as a result of burning out of the belt while a crown fire stops due to the absence of necessary support of a ground fire. Any retarding natural feature or obstacle, such as roads, paths, brooks and rivers, protective belts, stone deposits, small depressions with green grass, etc., to the spread of fire is used as a supporting belt. In case of necessity, artificial supporting belts are constructed manually by means of spades, manual sand casters, horse plows, solutions of fire-extinguishing chemicals or corded explosives. The width of the supporting belt depends on the expected dimensions of the flame and the intensity of combustion at the time of starting the backfire in relation to the reserves and character of the combustibles in the forest. It should be remembered that the supporting belt 'works' only for a few minutes, until the edge of the backfire moves away from it to a distance which prevents the transfer of the flame through the belt.

Manual creation of supporting belts, with sand casters and plows does not differ in principle from the technique of laying protective belts which have been examined in detail above. The difference between supporting and protective belts, lies only in the width of each. The width of a supporting belt rarely exceeds 60 cm, while the width of a protective belt is over l m.

A comparatively new method of creating a supporting belt is the use of a corded explosive charge for this purpose (Kurbatskii, Valendik, 1970). The corded charge consists of standard 200 g cartridges of ammonite *6-ZhV*, packed with end faces near one another with a gap of 3 mm and enclosed in capron braiding (cordage). The total length of the charge is 10 m, and the weight is nearly 9 kg. The charges in the form of coils are packed in

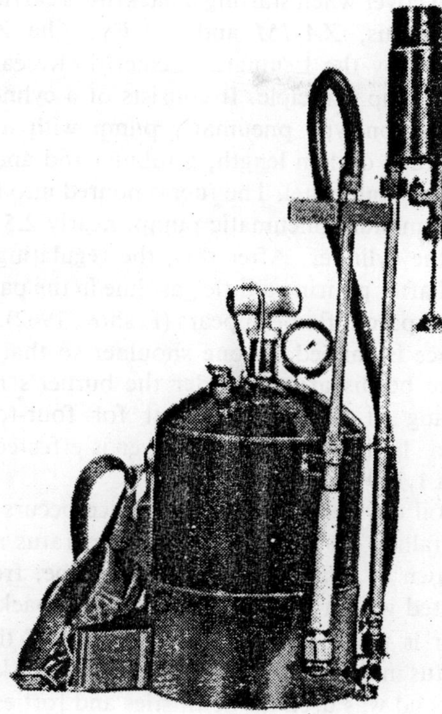

Fig. 33. Ignition device *ZA-1M* designed by the Leningrad Scientific Research Institute of Forestry.

polyethylene bags and then in jute bags, which protect them from dampness and are convenient for transporting them to a fire area in any type of mechanized transport. When transported by helicopter, they can be thrown out from a height of up to 50 m. On the track of the protective or supporting belt, they are laid out one after another closely pressed to the ground without defusing. The ends are joined with butt overlapping and secured with threads. Explosion is effected by means of a primer and a safety fuse. As a result of the explosion, a belt or trench is formed at the site of the corded charge, depending on the character of the top soil horizon. In fine skeletal soils, the width of the belt turns out to be 40 to 55 cm. In peaty soils, instead of a belt, it turns out to be a trench 30–60 cm deep and up to 1.5 m wide on the surface. On frozen ground with a thawing depth of 10–50 cm, the depth and width of the trench is slightly less.

The main feature of corded charges is that they help in laying out supporting belts or trenches in places, where due to soil conditions, or difficulty in transporting equipment, the use of other methods for liquidating fire is impossible.

Ignition of the soil cover when starting a backfire is carried out by means of the ignition apparatus, *ZA-1M* and *ZA-FK*. The *ZA-1M* ignition apparatus is developed by the Leningrad Scientific Research Institute of Forestry on the blow lamp principle. It consists of a cylinder of 7.5-liters capacity of fuel, a piston-type pneumatic pump with a manometer, a flexible rubber cloth tube of 1 m length, a tubular rod and a burner (Fig. 33). Kerosene is used as liquid fuel. The fuel is poured into the cylinder and then by means of the manual pneumatic pump, nearly 2.5 atm pressure is then created inside the cylinder. After this, the regulating control of the burner is opened and after pouring a little gasoline in the pan of the burner, it is heated until a transparent flame appears (Ershov, 1962).

The *ZA-1M* device is carried on one shoulder so that in case of dire necessity can at once be discarded. Under the burner's normal working conditions, one filling of fuel is sufficient for four-to-five hours of continuous operation. Ignition of the soil cover is effected by a flame jet with a temperature of 1,100–1,200°C.

Ignition of the soil cover by the *ZA-FK* device occurs due to burning droplets of the fuel falling on the cover as the apparatus moves along the belt. The fuel is drawn by gravity along flexible tube, from the cylinder toward the outlet fitted with a wick. When starting a backfire, the rate of ignition of the cover is much higher than with help of the *ZA-1M*. The weight of the apparatus in the filled state does not exceed 6 kilograms.

The backfire method was used in the thirties and forties for controlling large fires in the Urals and the northwestern and other regions of the USSR, and gave good results under favorable conditions (Serebrennikov, Matreninskii, 1937; Nesterov, 1945). Before starting a backfire, it was recommended to hack a fire lane and construct a wooden barrier or brakeline of man's height out of the resultant wood. On the approach of the fire front toward the barrier up to the distance where the draft of the oncoming air was felt, the brakeline was simultaneously ignited along its whole length. The strong flame of the burning brakeline rose to meet the fire. When the two flames came together, further advance of the fire front stopped. However, the use of such a method entails considerable effort in clearing a lane in a short time, forming a long line of combustibles and also entails a risk of forming a new source of crown fire as well as a greater danger to people living in the area.

V.G. Nesterov (1945) suggested the use of annealing or ground backfire against crown fires. He pointed out that on its approaching the fire front, ground fire is transformed into crown fire due to the draft and destroys it. This method of controlling crown fires, marked by its simplicity, high efficiency and comparative safety, is now widely used.

A disadvantage of ground backfire is its slow (not more than 0.6 m/min) advance against wind. In order to stop rapid crown fire, it is essential that the width of the burnt out belt should not be less than that of one 'jump' of the fire, i.e. 100–150 m. Thus, in order to construct a wide annealing belt, a backfire should be started several kilometers away from the fire front, which is organizationally complicated and therefore difficult to carry out. Constant crown fires can be stopped by ground backfire 40–50 m away from the supporting boundary. The main problem in this case lies in the liquidation of ignitions occurring beyond the supporting line, due to sparks. The width of the burnt out belt for stopping a ground fire front should not be less than 10 m.

For accelerating the process of burning out soil cover in the belt, N.N. Egorov (1955) suggested a several meter withdrawal from the supporting line on the side of the fire front some time after the start of the ground backfire, as well as igniting the cover along a line parallel to the supporting line. The speed of burning of the belt sharply increases with such a pattern, but the flame on the additional lines will advance against wind as well as with the wind, thus threatening to create a source of crown fire. Besides, a worker with an ignition apparatus who is present between the fire edge and the annealing line can easily be caught in a fire trap.

A more suitable method of accelerated belt burning before the fire front has been suggested by V.P. Molchanov. This method mainly consists in burning along lines perpendicular to the direction of the wind after burning out a narrow belt (1–2 m) from the supporting line. With this method, the burnt out belt develops slower than by the earlier variation, but entirely meets safety requirements and, therefore, is more often used than other methods.

Suppression of Underground (Peat) Fires

Heterogeneous flameless combustion is characteristic of underground (peat) fires during which nearly 50% of the liberated heat is used in drying and heating the boundary layer of the peat or forest floor. Peat fires are capable of self-penetration, which makes their control extremely difficult.

The principal methods for controlling peat fires include digging trenches up to the mineral soil along the perimeter of the fire area, and extinguishing with water. The first method is extremely laborious and unsatisfactory. When fire emerges on the wall of the trench, its transfer across the trench is not ruled out. After extinguishing with water, fires often revive, since water, due to its relatively high surface tention (72 dynes/cm), does not wet the particles of decomposed peat containing up to 18% of oily bitumens. Besides, a large quantity of water is required in order to wet nonburning peat layers up to a 400% humidity when combustion stops.

In 1963–1964, N.N. Krasavina (1963) and V.G. Lorberbaum (1963) developed an effective method of extinguishing peat and litter-humus fires at the Leningrad Scientific Research Institute of Forestry based on improving the wetting power of water with the addition of a small quantity (up to 0.3% in weight) of a surface-active substance. A small addition of a wetting agent sharply reduces the surface tension of water up to 34 dyn/cm, making it 'wet'. Sulfanol *NP-1* is most often used as the wetting agent although detergents *OP-7, OP-10, Progress* and others can be used with similar success.

Water is not simply poured on the fire, but is introduced into burning layers by means of a peat barrel specially constructed for this purpose. A peat barrel *TS-1* (see Fig. 26) is a brass tube 1.3 m in length and 16 mm in diameter, ending in a detachable cone. In the upper part of the tube, a regulator with a nozzle cap nut is soldered at a right angle for connecting the barrel to the hose line. The tube ends with a T-shaped handle. In the lower part of the tube, there are 40 holes extending over 40 cm, each 3 mm in diameter. Through them, water with the wetting agent enters the lower layers of peat after the barrel is lowered into it.

The technique of extinguishing peat fires consists of the following: a beginning is made with the liquidation of the ground fire edge, from where the peat fire usually starts. This method is the same as was described earlier,

Fig. 34. Suppression of deep-seated peat fires with a wetting agent *TS-1* solution through a peat barrel.

viz. extinguishing ground fires by means of a motor pump, hose line and fire barrels. After this, the limits of the deep-seated peat fire areas are established, which during the first days are estimated to cover tens of square meters, as well as the possible depth of the peat layer burn-out. It should be remembered here that it is often difficult to establish the underground boundary of a combustion zone only by external symptoms. Therefore all precautionary measures should be taken when approaching the open part of a fire area. Fire suppression is started from the wind-exposed, smokeless side of the fire area, since a peat fire does not have front, flanks or rear, and develops more or less uniformly on all sides. The barrel with the closed regulator is introduced with some force into the entire depth of the burn-out area and the regulator is opened. After waiting for some time, the regulator is closed, the barrel is again pulled out and, leaving a space of 30–40 cm along the edge, the barrel is again introduced, and so on, covering the fire area in a circle (Fig. 34). For complete liquidation of the fire, it is necessary to run the barrel over a second row of boreholes, arranging them at a distance of 30–40 cm from the first row in a checkerboard manner. The time of water delivery through the hole depends mainly on the depth of the peat burn-out. It is determined experimentally:

Depth of peat, cm	20–40	40–70	70–120	120–200
Time of fluid delivery, s	5–6	10–12	10–12	14–16

The recommended pressure in the hose line is from 3 to 6 atm. If the delivery of water with sulfanol from the tank of a road tanker, it is recommended that the above rates be maintained in order to avoid excessive consumption. Sulfanol is added into the tank filled with water (and not the other way round, since it results in the formation of foam) at the rate of 3 kg per ton of water. While the tanker moves toward the fire area, sulfanol dissolves completely in the water. Delivery of water with sulfanol can be carried out even from open water reservoirs by means of motor pumps *ML-100, M-600* and *MP-800.* In the motor pump *ML-100,* a rubber tubing, 1.5–2 m in length and 5 mm in internal diameter, is attached to the regulator of the gear pump. The free end of the tubing is immersed into the bucket with 15–20% solution of the surface-active substance. Water for the solution enters the bucket through a delivery tube of the same length and section as the previous tubing, the free end of which is also immersed in the bucket. The rate of water flow into the bucket and pumping of the solution from it are controlled by the tubing regulators. After this, during the operation of the pump, the only requirement is to add sulfanol powder at intervals. Fluid wetting agents of the *Progress* type can be lifted by the pump in an undiluted state. In order to obtain optimum concentration of the wetting agent in water (0.3%) at the end of the base line, the following flow rate of the concentrated solution, depending on the type of chemical and pressure developed by the pump, is recommended:

Pressure, atm	3–4	5–6
Rate of wetting agents, kg/h of sulfanol *NP-1* (diluted with water 1:5)	5	7
OP-7 or *OP-10* (diluted with water 1:1)	10	12
Progress (not diluted with water)	15	20

The complete set of equipment for motor pumps *M-600* and *MP-800* should include two insertion pieces with regulators for suction and discharge nozzles. The insertions are simple in construction and can be manufactured in any forestry or wood industrial mechanical workshop. Intake and delivery tubes are attached to the regulators, and the technique of water delivery with the wetting agent is the same as with the operation of motor pump *ML-100* Fires, which have penetrated less than 1.2 m, can be extinguished with peat barrel *TS-1*. When extinguishing deeper peat layers, barrel *TS-2* of 2 m length is used. In the latter modification, the regulator is built into the handle.

Large-scale commercial trials of the above method of extinguishing peat fires in the territory of Gorki province have shown its high efficiency and reliability. Of the 60 fires extinguished with a burn-out depth of peat of up to 130 cm, there was not a single revival.

Aerial Methods of Suppressing Forest Fires

In thickly forested and sparsely populated regions of the north, Siberia, and the Far-Eastern part of the USSR, the use of aircraft is the principal means of detecting forest fires and for quick transport of men and fire-fighting aids to these areas. Bases for air protection of forests carry out fire-fighting operations in an area of over 700×10^6 hectares, which is nearly 80% of all the forests in the country.

With modern technical equipment, the air bases perform two major functions: aerial patrolling and active fire-fighting. Aerial patrolling involves regular flights over a particular region for the purpose of early detection of forest fires, precise determination of their locations and conditions of their spread, and also notifying the forest ground protection service.

Active control of fires is carried out in regions where their liquidation by ground forces cannot be ensured. Direct suppression of fires is carried out by fire parachutists and airborne brigades equipped with light fire equipment (knapsack sprayer, spades, ignition apparatus) and explosives. *AN-2* aircraft are used for landing fire parachutists in a forest fire area, while for landing airborne fire brigades, helicopters *MI-1, MI-4, KA-26, MI-2, MI-8* and *MI-6* with a carrying capacity of 0.3 to 10 tons are used.

The fire parachutist service is the most sophisticated in relatively populated regions, maintained by the forest air protection forces. Under these conditions, the parachutists quickly return to their base using local transport routes after liquidating a fire in order to be ready for the next jump. In remote, almost inaccessible regions dominated by overmatured forest stands with many dead standing trees and also in mountainous regions with exposed rocks, stone deposits, etc., the dropping of fire parachutists is hazardous. In these regions, the transportation of workers and fire-fighting equipment is carried out by helicopters. Light helicopters of types *MI-1* and *KA-26* are used mainly for protection of areas, where small numbers of fires occur, while medium and heavy helicopters (*MI-2, MI-4, MI-6, MI-8*) are used in regions with a higher fire risk.

In recent times, the efficiency of the fire parachute service has greatly improved due to its operational capability and technical equipment. New types of parachutes have appeared which enable precision landing on small forest airfields, and special clothing is available which protects parachutists

Fig. 35. Fire parachutists in protective suits ready for jumping in forests.

jumping directly into the forest (Fig. 35).

The jumping of workers of airborne brigades from helicopters is effected by means of lowering devices, with helicopters hovering close to the fire area. Due to the operational capability, and skilled training of the personnel of the parachute-landing service as well as the operational capability of aerial forest protection service in the zone, over 85% of all detected fires are liquidated jointly by the air bases and fire protection service personnel.

However, the light fire equipment supplied to the fire brigades is not very effective when fighting extensive fires, while transport of heavy equipment is limited by the load-carrying capacity of most helicopters and the transportability of the equipment itself to inaccessible regions. Therefore, the use of planes and helicopters for the direct suppression on fire from the air should be considered the chief target in the further improvement of forest air protection.

The first experiments on forest fire suppression from the air were conducted in the country in the forties under the guidance of P.P. Serebrennikov (1937). Water was discharged from light planes, *PO-2* and *R-5*, which took small supplies of fluid on board. Due to this reason, no positive results were obtained, and operations were stopped.

In 1950, *TsNIILKh**, *GosNIIGVF*** and the design office of O.K.

**TsNIILKh*—Tsentral'nyi nauchno-issledo-vatel'skii institut lesnogo khozyaistva (Central Scientific Research Institute of Forestry)—Translator.

***GosNIIGVF*—Gosudarstvennyi nauchno-issledo-vatel'skii institut grazhanskogo vozdushnogo flota SSSR (State Scientific Research Institute of the Civil Air Fleet USSR)—Translator.

Antonov developed an aerial fire sprayer-*APO,* to be fitted on an *AN-* aircraft. Fluid from a 1,200-liter tank was ejected under pressure through special nozzles in a 'backward' direction to the course of the plane at a speed of 40 m/s, i.e. about the same as the flying speed of the plane. With such a method of outflow, breakup of the fluid spray was less and evaporation losses were considerably reduced. With a flying speed of 140 km/h and a height of 35–40 m, it was possible to lay out protective belts of 8–13 m in width and up to 160 m in length under a canopy of mature pine stands. Construction of the *APO* was an important technical improvement in solving a complex problem. However, this apparatus did not find an application in aerial forest protection operations. Firstly, the fluid tank and the compressed air bottles for forced ejection blocked up the cabin of the plane and created an inconvenience during patrolling with firemen-parachutists on board. Besides, it was necessary to return to the airfield for the next filling of fire-extinguishing fluid which sharply reduced efficiency of operation. And, finally in spite of the complex nature of the water discharging device, it was far from perfect.

During the fifties and sixties, experimental and design work directed toward solving the problem of extinguishing fires from an aircraft was conducted more intensively in Canada and the US. As early as 1944, Canadian specialists arrived at the conclusion that under conditions of an extended network of rivers and lakes, it is advisable to use seaplanes capable of removing water from the nearest weir (Tamarkin, 1966). It is sufficient to mention that in the Canadian province of Ontario alone, there are over 200 thousand lakes. A large number of natural water reservoirs suitable for landing of seaplanes are to be found in forest regions of Quebec and British Columbia.

Over the last 20 years, seaplanes and airplanes of more than 10 different types with a carrying capacity of 0.5 to 27 t fitted with forest fire equipment were manufactured and tried out in Canada and the USA.

The first experiments were conducted in 1944 in Canada with the seaplane, *Northern Aircraft,* the tanks of which were filled with water while on stand-by duty. But results were unsatisfactory due to the low carrying capacity of the plane and the slow intake and discharge of water. After two years, special projectiles filled with water which were to be dropped from heavy bombers were tried for extinguishing forest fires in the State of Florida. These results were also found to be unsatisfactory and, moreover, the projectiles themselves created a risk for people present in the fire area. In 1950, suppression of experimental fires by dropping bags filled with water was tried in Ontario. In spite of individual successful hits, the specialists conducting these experiments made adverse comments.

The first favorable results were obtained in 1953 by one of the American aircraft companies, after conducting experiments on the effect of the outflow of water on fire from *DS-7* planes with a carrying capacity of 5,000 liters. Water outflow from a height of the order of 130–150 m at a speed of 200 km/h resulted in the formation of a wet protection belt of 60 × 160 m on the ground. These experiments gave a stimulus to further experiments on extinguishing fires from the air.

In 1956, seven *Styrian* planes with a carrying capacity of nearly 5,000 liters were used for extinguishing 25 fires, while in 1957, seven *Avenger* planes, two *Canso amphibian* planes and one *Fairchild Packet* plane were used. From this time, in spite of isolated failures, the method of extinguishing forest fires from the air began to be used more and more extensively in the USA and especially in Canada, where due to a poor network of roads in the northern regions of the country, it was difficult to employ other methods of fire control.

Practice has shown that the simplest and most effective method of extinguishing fires is the use of a free flow of water, or water with the addition of fire-extinguishing chemicals.

According to Canadian and American specialists, a tanker plane is able to carry out the following main tasks: to land personnel of the brigade near a fire area with fire fighting equipment, to delay development and spread of the fire until the arrival of ground forces and to fight the fire in the most dangerous directions. The last task can be carried out by cooling the most intense burning areas, constructing obstructive barriers with chemical solutions, preventing transformation of a ground into a crown fire and by increasing the fire-resistance of adjacent forest areas (Tower, Wally, 1971).

In the USSR, the first favorable results in extinguishing forest fires from a seaplane were obtained in 1964 in the Tyumensk province, where the forest fire model of the *AN-2V* (*AN-2P*), developed by the design bureau of O.K. Antonov according to technical specifications of the Leningrad Scientific Research Institute of Forestry, *GosNIIGA** and the Central Air Base, underwent trials under the guidance of V.P. Molchanov. Floats with compartments (Fig. 36) were used in an *AN-2P* plane as water tanks. The lower part of the water-intake compartment has two hatch flaps, viz. for intake and outflow. The intake flap is fixed on the side of the floats which is external to axis of the plane, while the other is on the internal side. The capacity of each compartment is 630 liters. There is provision for filling in quantities of 300, 500 and 630 liters. Filling of the compartment is effected in five-to-seven seconds as the plane glides over a water surface. For intensifying the fire-extinguishing properties of water, after intake, a portion of a concentrated solution of a wetting agent is added to it in the compartments. For such supply of the wetting agent, a special automatic system has been set up which consists of two tanks of 50 liters each, dosed cans of 5 liters each, overlapping regulators and pipes. Control of the water

**GosNIIGA*—Gosudarstvennyi nauchno-issledo-vatel'skii institut grazhdanskoi aviatsii (State Scientific Research Institute of Civil Aviation)—Translator.

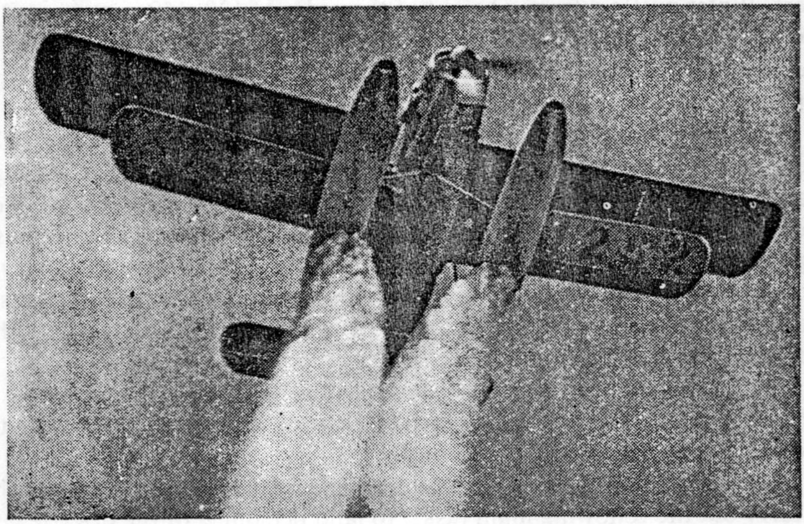

Fig. 36. Forest-fire seaplane *AN-2P*.

intake process, addition of the wetting agent, and the outflow of water during flight are automatically controlled from the cockpit.

At the end of the experimental-commercial trial of the *AN-2P* plane, it was established that it had a high efficiency in extinguishing forest ground fires in pine stands of 22 m height and up to 0.8 density, and also in thin forests and open areas. It was also established that wet belt of up to 100 m in length (average 70 m) and 8–12 m in width is formed by one water outflow. Fire-fighting belts, laid out from the air, extinguish a flame edge, or sharply reduce intensity of combustion. With workers available at the fire site, the fire edge can be easily suppressed and liquidated from the air. However, in the absence of firemen, it will gradually start reviving after some time. The revival rate of flame at the fire edge after aerial treatment depends mainly on the type of forest and its density. In lichen-covered forests with low and medium density (20–30 m^3/ha), the fire edge revives after 35–40 minutes, but with high density and in young woods, combustion revives after 5–10 minutes. The time interval is the same for the revival of the edge in ledum-covered pine forests, irrespective of their density. However, due to comparatively low engine power, the *AN-2P* plane did not find wide application in aerial forest protection operations in our country.

Essentially new types of aircraft, viz. helicopters, first began to be used in aerial forest protection operations in the country from 1955, and at present they constitute over 60% of the total number of aircraft being used by the air forest protection service. This is explained by the fact that they have several advantages over airplanes in carrying out certain types of operations, for example, the transport of personnel and loads to the fire area and their return journey.

At present, *MI-1, MI-2, MI-4, MI-8, KA-26* helicopters and periodically *MI-6* are used by the air forest protection service (Biryulin, Makarov, Kanishev, 1969).

The effectiveness of helicopters in forest protection largely arises from their flight performance (Table 20).

TABLE 20. FLIGHT PERFORMANCE OF HELICOPTERS USED IN AIR FOREST PROTECTION SERVICE

Flight performance specifications of helicopters	Types of helicopters					
	MI-1	*MI-2*	*MI-4*	*MI-6*	*MI-8*	*KA-26*
Maximum takeoff weight, kg	2470	3550	7350	42500	12000	3480
Weight of structure, kg	1911	2350	5495	27200	7500	2022
Maximum payload, kg	380	700	1625	12000	4000	1035
Cruising speed, km/h	130	205	150	250	200	140
Service ceiling, m	3010	4000	5500	4500	4000	31000
Maximum flying range, km	422	597	650	810	640	410
Number of engines, *items*	1	2	1	2	2	2
Maximum engine, hp	575	400	1700	5500	1500	325

In 1957, for the first time, experiments on extinguishing forest fires from helicopters were conducted in the country by *TsNIILKh, GosNIIGVF* and the Central Air Base. Easily detachable forest fire equipment for the *MI-4* helicopter included two tanks of 400-liters capacity each, the motor pump *M-800* and a special barrel. A small-sized motor pump was used for filling the tanks with water. The results of trials have established the possibility of laying out protective belts from the air of up to 120–150 m in length and two-to-four mm in width in felling areas (Fig. 37), in young woods and under the canopy of mature pine spruce stands. However, due to the limited efficacy of such belts, their short length for each outflow, the necessity of landing near a reservoir for water intake and the comparatively high cost of operation of *MI-4* helicopters, the results of experimental trials on this method of controlling fires showed that it is uneconomical.

Later, in 1966, operations for extinguishing forest fires from the air were continued with the *MI-6* helicopter. This helicopter was equipped with a flexible tank of 4 m^3 capacity placed in the cargo hold. The tank had a filling hole, drainage hole and an outflow hose, 700 mm in diameter and 1.5 m in length. The tank was filled by means of a small pump, while the outflow was effected by gravity. Outflow of the water over open areas from a flight altitude of 10–15 m and speed of 50–60 km/h resulted in the formation of wet belt on the land surface of nearly 100 m in length and 6 m in width, with fluid doses of 60 to 0.2 liter per square meter. With an increase in altitude and also in flying speed, the width and length of the belts

Fig. 37. Construction of a barrier belt from a *MI-4* helicopter.

also increased with a corresponding reduction in the doses of the fire extinguishing fluid. However, a disadvantage of the equipment was the necessity of choosing from the air, a suitable place for landing near a reservoir for filling, which is not always possible under taiga conditions.

In 1967, forest fire equipment was developed in the USSR, which permitted water collection from a reservoir by a *MI-6* helicopter hovering over the surface. The complete set of the equipment includes: a nylon tank of 12,000-liters capacity with a two-meter appendix hose of 650 mm diameter, 5 pumps with a delivery rate of 60 m^3/h and 4 barrels fixed on a special mount. Water intake from the reservoir was effected on a low level flight over the water surface by means of two rods 41 m length, at the ends of which high power electric pumps are mounted. Protective belts were created by two methods: by free discharge and by the delivery of water under pressure by means of pumping devices. Water was released directly on the fire edge at a flying altitude of 20 m and a speed of 20 km/h. One discharge to the ground resulted in the formation of a wet belt 18 m in width and 200 m in length. With a distance of less than 5 km between the water reservoir and the fire area, the entire cycle of intake-release-intake took about 4.5 minutes.

In the latest modification, the fire equipment of the helicopter has been supplemented by a fire-fighting monitor installed in the front part of the cabin, and a reservoir for a foam compound.

In Canada and the US, experiments were conducted on the use of helicopters for laying fire hoses, for extinguishing fires from the air by discharging fire-extinguishing fluid in special bags and by discharge from tanks suspended externally. Special tanks (*Canadian barrels*), transported on the external suspender of the helicopter (Wajax, 1972), found widest application in extinguishing forest fires in these countries. These are small cisterns manufactured from fiber glass plastics and controlled by mechanical power cable devices. The capacity of such cisterns varies, depending on the load-carrying capacity of the helicopter, from 250 to 3,000 liters. Water collection is done by ordinary scooping from natural or artificial water reservoirs and even by tipping bodies as well as by means of special valves in the bottom of the tank, which open due to water pressure while lowering the tank in water and close with the pressure of collected water. Water discharge or discharge of fire-extinguishing fluid is carried out by inverting the tanks during flight over the fire edge or through a valve at the bottom of the tank. A very important advantage of such devices is the simplicity of design and operation, and their low cost. With the availability of water sources near the fire, the use of such simple tanks ensures high performance of the helicopters in the conveyance and discharge of water. Thus, during three days of a fire liquidating operation in the National Forest of the State of Oregon, one helicopter poured out 55.6 t of water and 98,400 liters of thickening agent.

The presence of suspended tanks enables the use of standard helicopters for extinguishing a fire without any structural modification or complex assembly equipment.

In recent years, helicopters fitted with the most modern equipment began to be used in Canada for extinguishing forest fires. Thus, a tank of 340 liters capacity manufactured from water-proof corrugated material in an aluminum framework is secured at the bottom of the sea model, 'Dominion' helicopter.

The tank is filled with water by means of two electric pumps with a total delivery rate of 4.5 liters per second after the helicopter lands on the water. Discharge of water or chemicals is effected through an elastic vent. The system of intake and discharge of water is completely automatic. In case of failure of the automatic system, there is a provision for manual control. Depending on the flying speed, the length of the wet belt can be from 1.5 to 9 m.

At present, work is being conducted in the country on the construction

Fig. 38. Water discharge equipment for a *KA-26* helicopter.

of special equipment for standard helicopters designed for extinguishing fires from the air. It has found wide application in aerial forest protection. Interest in the development of fire equipment for helicopters has grown especially due to the introduction of new and more better equipped helicopters of the *KA-26* and *MI-8* type in aerial forest protection. Thus, a device was developed and manufactured in 1971 for intake, transport on the external suspender, and discharge of water from a *KA-26* helicopter (Fig. 38) (Artsybashev, and others 1972). It includes a tank and suspension system, the tank being made of glass fiber plastic. The capacity of the tank is 320 and 420 liters (regulated by means of four plugs); the weight is 23.6 kg. In a stowed position, the tank is secured outside the freight and passenger cabin of the helicopter on rigid mounts. The external suspender of the tank consists of two tackles, 10 m in length, connected to the cabin walls and secured on the beams under the ceiling of the freight and passenger cabin of the helicopter. In the lower parts of the tackles, there are spring hooks, allowing quick coupling and uncoupling of the tank. In the upper part of the external suspender under the ceiling of the cabin, a special device—a *cable cutter* is provided in case of sudden uncoupling of the suspension system and the tank in emergency situations. A switch in the cockpit serves for controlling the intake and discharge of water.

During trials of the water discharging device, parameters were determined for the belt on the soil surface that is effectively moistened. A belt with a dose of not less than 2 liters per square meter should be considered as an optimum protection belt. Such a dose of fire-extinguishing fluid is considered sufficient for stopping the passage of the edge of a ground fire of average intensity. With experimental discharges in an open area, an effective wet belt had a length of 30–35 m and a width of 5–7 m. With water discharges from an altitude of 10 m above the crowns and a speed of 20 km/h over pine stands of 20 m height and density of 0.6, the length of the moist belt did not exceed 24–28 m with a width of 4–5 m. It was established that during discharge, nearly 20% of the water evaporates in the atmosphere and does not reach the ground. Besides, some quantity of water (nearly 12–14%) is retained by tree crowns.

Doses of water on the surface of forest soil cover vary substantially. Thus, in the 'epicenter' of the fall of the main mass of water, a high concentration (up to 5–6 liters per square meter) is observed, while in peripheral areas, the dose is sharply reduced and constitutes about 0.2–0.5 liters per square meter. Nonuniformity of wetting in a forest should be taken into consideration when extinguishing fires by means of this device. In the opinion of the authors, suppression of ground fires from a *KA-26* is not advisable, since the moistened belt in most cases will be ineffective, when the density of the forest stand is 0.6 above and the height of the trees, 20 m or more. Under such situations, best advantage can be derived from a helicopter, if it is used for transporting water to the fire area for supply to ground fire fighting groups.

Experience at home and abroad shows that extinguishing from the air alone without the support of ground forces rarely succeeds in liquidating a fire. An airplane or helicopter should observe the fire until the approach of elements of the freshly landed fire brigade and then help the ground forces from the air until final liquidation.

The load-carrying capacity of aircraft tankers is a very important index of the efficiency of their application. A plane of low carrying capacity can take in water from small rivers and lakes and discharge it from a low altitude, thereby reducing losses by evaporation in the atmosphere. The cycle of discharge-intake-discharge takes not more than three-to-four minutes, if the distance between the fire and the reservoir is not more than 5 km. Side by side with extinguishing fires, such a plane can also carry out patrolling.

A disadvantage of planes with low carrying capacity is the impossibility of achieving reliable results in extinguishing a fire edge in one sortie. Heavy planes ensure creation of more effective protective belts, but they are less maneuverable, require large expenses for the intake of water, and it is disadvantageous to use them for patrolling. Besides, a large mass of water discharged from a low altitude may turn out to be dangerous for people on the ground.

On analyzing the data of using aircrafts for extinguishing fires in the country and abroad, it can be concluded that for regions with a large number of lakes (over 10 for 1,000 km²) and rivers, a plane with a carrying capacity of about 2 t should be considered as most effective for meeting the requirements of an aerial forest protection service. With the number of lakes varying from 1 to 10 over an area of 1,000 km², the capacity of a plane in extinguishing fires is sharply reduced due to an increase in the average distance between the fire area and the reservoir. A plane with a carrying capacity of about 5 t should be considered most suitable under these conditions.

The capacity of a plane or a helicopter to extinguish fires from the air can be expressed by the number of water discharges which it can carry out per unit of time and the length of the suppressed fire edge, under the condition that the final liquidation of the fire is to be accomplished by workers present at the site.

The number of water discharges per hour depends on the distance between the fire area and the reservoir and can be determined by the formula

$$N=\frac{60}{\frac{120L}{v}+t},$$

where L is the distance between the fire area and the reservoir in km, v the velocity of the plane or the helicopter in km/h, t the time spent by the aircraft for maneuvering during water intake from the reservoir and preparing to turn for discharge in min, $120L/v$ the time spent on the flight from the reservoir to the burning area and back.

The capacity of a plane or a helicopter to extinguish a fire edge can be found by the formula

$$P=NlK,$$

where P is the capacity of the plane or helicopter, under the condition that the final liquidation of the fire is achieved on the ground by personnel of the fire brigade, in m/h, l the length of fire edge which can be extinguished in one water discharge from the air, in m, and K the coefficient of impact accuracy (0.7), established experimentally.

Because of limited visibility, people on the ground cannot always understand the situation in the fire area, which can rapidly change depending on the character of combustion.

From a plane or helicopter engaged in extinguishing a fire from the air, it is not difficult to decipher the character of fire spread and estimate the tactical situation.

Fig. 39. Fire-fighting sound amplifier set, *PZS-68*, for patrol planes and helicopters.

For directing fire-extinguishing operations from the air, a special airborne sound amplifier set, *PZS-68,* has been evolved (Artsybashev, Orlov, 1970). The set consists of two main units, viz. the amplifier and the loudspeaker (Fig. 39). The amplifier is fully transistorized and manufactured in the form of a 628 × 415 × 337 mm portable suitcase and weighs 18 kg. The loudspeaker horn is 470 mm in height and 15 kg in weight. The nominal output of the amplifier is 75 W. It is powered by the aircraft electrical system. The loudspeaker horn is secured by means of a simple bracket in the hatch of the sprayer of an *AN-2* plane or outside the fuselages of *MI-1, MI-2, MI-4* or *KA-26* helicopters. Transmission of any information, orders or instructions are broadcast by means of special noise protected laryngophones, which ensure clear sound at all levels of engine noise. The best audibility on the ground is achieved with the plane flying at an altitude of 300 to 600 m. Good audibility time under these conditions is 30–40 seconds, which is sufficient for transmitting three-to-four phrases of five-to-seven words each. When broadcasting from a helicopter, the best audibility effect is achieved at the same altitude with the helicopter hovering. With an increase in the flying altitude of airplanes or helicopters, the audibility time at a selected point increases, but the sound intensity is markedly reduced.

Experience of using sound amplifier set *PZS-68*, by air base operational squads has shown that it can be successfully used for ensuring fire safety among the local population and especially among people present in the forest, for warning fire safety violaters, for messages about fire situations in forestry and forest administration bureaus, for training fire parachutists, for the search of people lost in a forest and so on.

Suppression of Massive Forest Fires by Artificially Induced Rainfall

After prolonged droughty periods, massive outbreaks of fire often occur in the thickly forested regions of the north, Siberia and the Far-Eastern part of the USSR, most often as a result of thunderbolts. In such a situation, some of the fires cannot be detected and extinguished in time, and they can engulf large areas. The number of such fires is small, but is precisely these which determine the inflammability of our forests forming part of the areas covered by aerial protection. During certain years, 90% of such burnt-out areas are due to large forest fires (Artsybashev, 1973).

It has been shown that enlistment of a large number of people and equipment does not always provide the desired effects. With great effort, it is sometimes possible to retard the fire edge in some directions, but more often large fires continue to be active for several weeks until the onset of the rainy season.

The great damage caused to the national economy by large forest fires necessitates a search for new solutions to this problem; and primarily involves developing means and methods for their suppression.

As a result of the efforts of meteorological and forestry scientists, an essentially new method of controlling large fires has been developed, viz. rainfall induced artificially from some forms of convective clouds (Artsybashev, Gubin, 1970).

During the fire risk period, the main cause of cumulus formation is vertical convective currents of air heated due to contact with the hottest parts of the ground (felling areas, burnt clearings, southern slopes, etc.). Rainfall in summer, occurs in temperate latitudes, when in addition to droplets, ice crystals appear in the upper supercooled part of the cloud. Because of the difference in vapor tension on the surface of a droplet and a crystal, the latter grows on account of the 'transfer' of moisture from the surface of the droplets and on falling into the warmer layers of the cloud, the crystals melt and are transformed into rain drops.

During hot summer days, it is possible to observe the formation of huge cumulus clouds over forest areas, in which all necessary conditions for the formation of rain develop, although these conditions are still not critical. In such a situation, a slight stimulus is sufficient (introduction of an ice-forming substance into the cloud in order to induce a heavy shower).

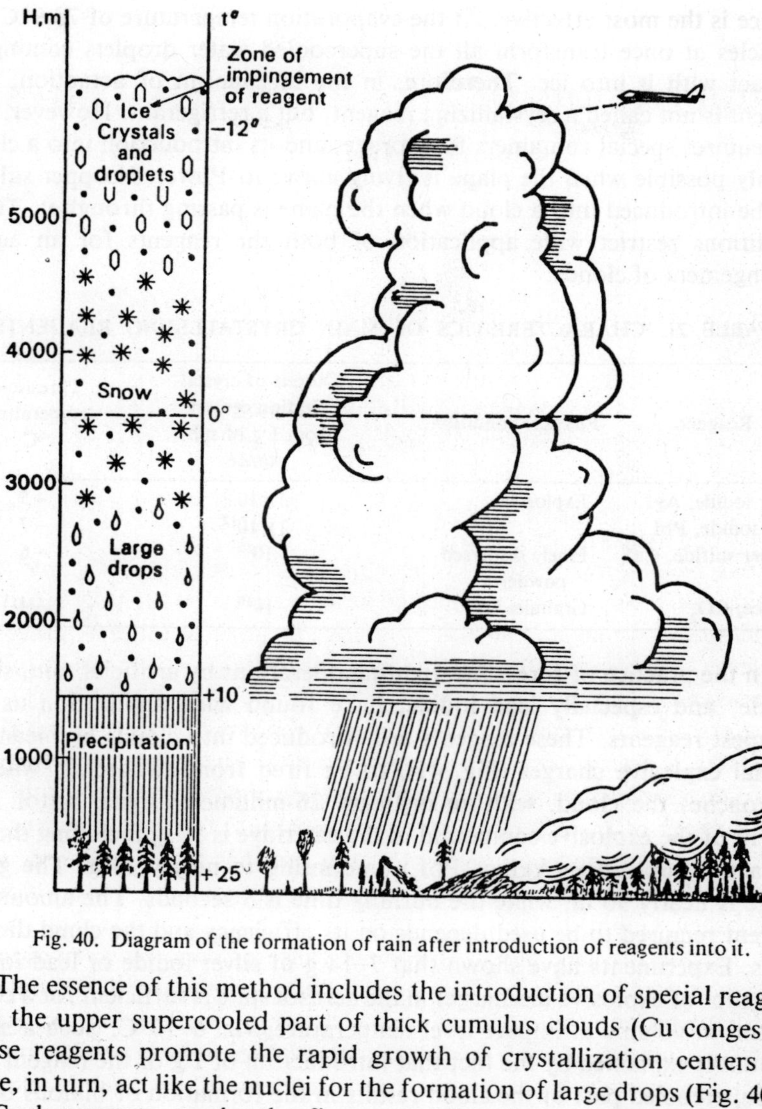

Fig. 40. Diagram of the formation of rain after introduction of reagents into it.

The essence of this method includes the introduction of special reagents into the upper supercooled part of thick cumulus clouds (Cu congestus). These reagents promote the rapid growth of crystallization centers and these, in turn, act like the nuclei for the formation of large drops (Fig. 40).

Such reagents can be the finest particles of silver iodide (AgI), lead iodide (PbI_2), copper sulfide (CuS) and also dry ice (CO_2).

The chief properties of a reagent, which characterize its intensity, are the output of crystallization centers from 1 g of the substance and its threshold temperature, i.e. the minimum temperature, under which crystallization starts in the supercooled part of the cloud after introduction of the reagent. The main characteristics of reagents, which are most widely used in the practice of active impingement of clouds, are given in Table 21.

As regards threshold temperature and output of crystallization centers, dry ice is the most effective. At the evaporation temperature of 78.9°C, its particles at once transform all the supercooled water droplets coming in contact with it into ice. Therefore, in the mechanism of actuation, this reagent is not called a crystallizing reagent, but a refrigerant. However, dry ice requires special containers for storage, and its introduction into a cloud is only possible when the plane is flying above it. Powered copper sulfide can be introduced into a cloud when the plane is passing through it. These conditions restrict wide application of both the reagents for an active impingement of clouds.

TABLE 21. CHARACTERISTICS OF MAIN CRYSTALLIZING REAGENTS

Reagent	Physical condition	Output of crystallization centers from 1 g of substance	Threshold temperature, °C
Silver iodide, AgI	Explosive	2×10^{15}	−5
Lead iodide, PbI_2	,,	2×10^{12}	−7
Copper sulfide, CuS	Finely dispersed powder	10^{10}	−6
Dry ice, CO_2	Granules	10^{16}	−4

In the practice of forest fire-fighting operations by artificial rain, silver iodide, and especially lead iodide, have found wide application as the cheapest reagents. These reagents are introduced into clouds by means of special explosive charges (Fig. 41), to be fired from an aircraft when it approaches the cloud, with an ordinary 26-millimeter signal pistol. The weight of the explosive compound in the cartridge is 30 g; including that of the active substance (AgI or PbI_2) amounting to nearly 15 g. The grain range is nearly 90 m, while the burning time is 5 seconds. The amount of reagent required to be used depends on its efficiency and the cloud dimensions. Experiments have shown that 7–14 g of silver iodide or lead iodide (one shot of explosive discharge) in the aerosol state is sufficient for a cloud 10 km^3 in volume at impact level temperature, viz. 6–10°C. Such a small amount is explained by the fact that introduction of 1 g of the reagent into the supercooled part of the cloud results in the formation of billions of ice particles, i.e. centers of moisture condensation.

Thick cumulus clouds of the Cu congestus type, moving in the direction of fire (Fig. 42), are selected for impingement. The direction of movement of a cloud can be determined by the direction of movement of its shadow on the ground.

Artificially induced rain from well-developed cumulus clouds begins to fall 10–12 minutes after the introduction of the reagent it, reaching the maximum after 20–40 minutes and continuing on the average for 50–70 minutes. The rainfall zone expansion depends on the thickness of the cloud,

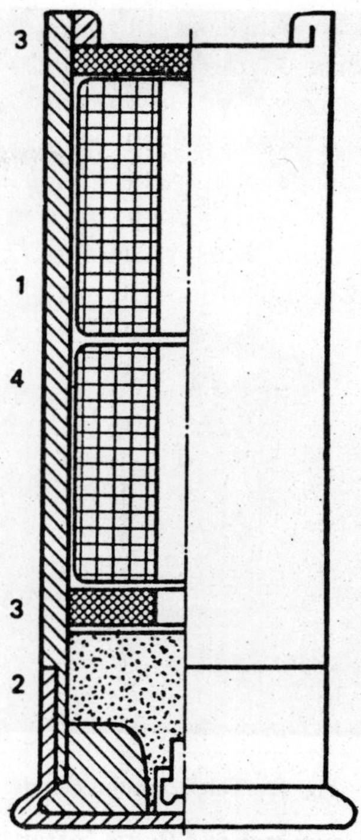

Fig. 41. Cross section of a 26 mm explosive cartridge for active impingement of clouds:

1—cartridge case; *2*—ejection charge; *3*—closing plugs; *4*—grain of explosive compound with reagent, AgI or PbI_2.

the width of its base and the drift velocity in relation to the ground. For conditions in eastern Siberia, the length of the belt moistened by the rain varies within a wide range—3 to 30 km, while the width varies between 2 to 6 km. For the Angara River bank conditions, the average volume of a cumulus with a thickness of 2.5–3 km constitutes about 9–10 km^3.

If the average water content of a thick cumulus is taken as 1 g of water per cubic meter of the cloud mass, precipitation of $(9–10) \times 10^3$ t in the form of rain can be expected from it. However, experimental measurements have shown that after the initial impact on the cloud, several times more rain is precipitated than the theoretically determined quantity. This is explained by the fact that in the process of artificial rain irrigation, the cumulus acts as an original generator, drawing in vaporous moisture from the surrounding atmosphere. After condensation in the upper part of the cloud, this moisture precipitates in the form of rain. According to preliminary data of G.F. Prikhot'ko (1968), the regeneration coefficient of a thick cumulus for

Fig. 42. Thick cumulus clouds.

Ukrainian conditions in summer is 4.2, i.e. after the impact, the rain yield of the cloud is about 4.2 times more than the initial reserve of water in it. To ensure higher probability of rainfall in the fire zone, it is essential of 'plant' with the reagent, not one cloud but several of them, encircling the fire zone like a 'horse shoe'.

In extinguishing forest fires with artificial rainfall, it is important to establish the criteria which determine the probability of rainfall after impingement of the reagent on the cloud. One of the most important criteria is the vertical thickness of the cloud, i.e. the difference between the height of the upper and the lower limits. Experiments conducted during 1969 to 1971 confirm this dependence (Fig. 43). All other conditions being equal, the probability of rainfall increases in direct proportion to the thickness of the cloud.

For estimating the efficiency and prospects of artificially induced rainfall, as a method of extinguishing forest fires, it is important to know how often cloudy conditions favorable for inducing rain develop during the fire risk period. The recurrence of days with resourceful (i.e. suitable for impact) cloudiness, is the most important characteristic in applying this method of extinguishing fires and helps in earmarking regions suitable for its introduction.

Data on the distribution of the number of days with thick cumulus and cumulo-nimbus clouds, according to the months of the fire risk period, are

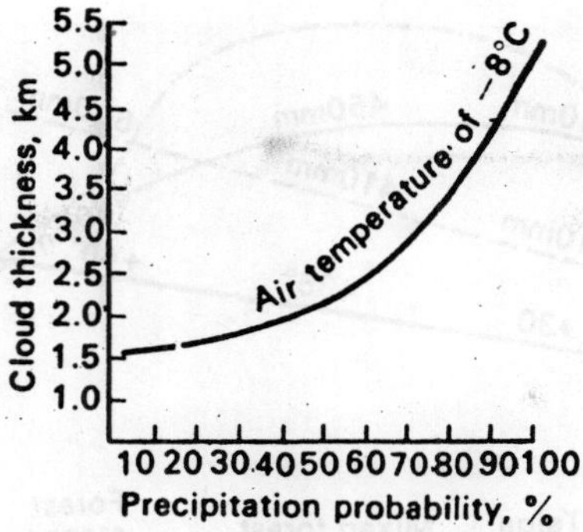

Fig. 43. Probability of rainfall as a function of the thickness of the cloud.

given in Table 22. The highest number of days suitable for impact are noted in the regions of Omsk, Krasnoyarsk, Kirensk, Aldansk and Khabarovsk meteorological stations. Tyumensk, Kolpashevsk, Irkutsk and Chitinsk meteorological stations and regions adjacent to Komsomol'sk on-Amur occupy an average position. And, finally, the regions of Khanty-Mansiisk, Novosibirsk, Yakutsk and Magadansk should be considered as less responsive.

The most favorable situation for using the method of active impingement of clouds develops in June and July, although in several regions (Tyumen', Khantymansiisk, Omsk, Kolpashevsk) it is also observed

TABLE 22. NUMBER OF DAYS WITH CLOUDINESS SUITABLE FOR IMPINGEMENT ACCORDING TO THE MONTHS OF THE FIRE RISK SEASON (AVERAGE DATA FOR 1961–65)

Meteorological station	Number of days with prospects of cloudiness					
	May	Jun	Jul	Aug	for 4 months	%
Tyumensk	6.3	9.0	11.6	11.7	38.6	31
Khanty-Mansiisk	4.7	5.9	6.8	10.1	27.5	22
Omsk	9.0	10.4	11.6	13.4	44.4	36
Kolpashevsk	5.9	8.1	8.5	9.5	32.0	26
Novosibirsk	2.6	5.4	4.6	5.4	18.0	15
Krasnoyarsk	7.8	12.6	15.0	12.0	47.4	38
Irkutsk	5.6	8.2	10.6	7.8	32.2	26
Kirensk	16.2	13.0	10.2	8.6	48.0	39
Yakutsk	6.6	8.0	6.6	4.4	25.6	21
Aldansk	3.6	16.4	12.4	11.8	44.2	36
Chitinsk	5.4	9.4	10.8	9.0	34.6	28
Komsomol'sk-on-Amur	10.6	10.6	6.0	8.0	35.2	29
Khabarovsk	9.6	11.8	10.8	10.8	43.0	35
Magadansk	—	0.8	2.4	0.4	3.6	3

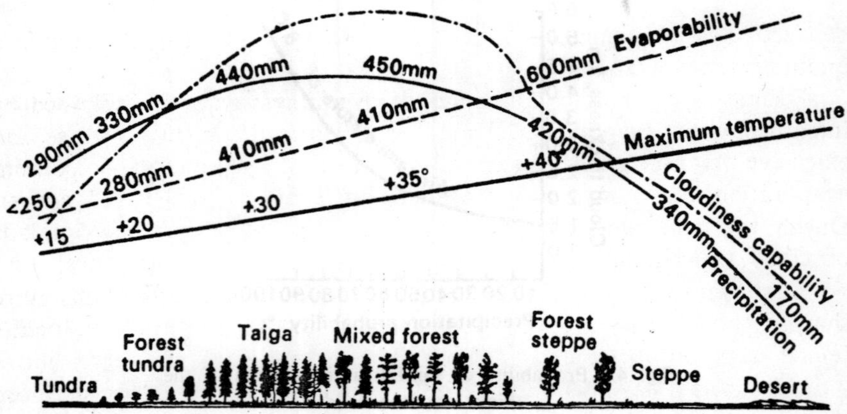

Fig. 44. Diagram of the meteorological profile through the vegetative zones of the USSR.

in August. In May, when the ground surface is still insufficiently heated, thick convective clouds rarely appear.

Cloudiness suitable for impingement is created as a result of circulatory processes having a definite directional effect during summer and also due to the effect of the underlying surface. Vast territories of western and eastern Siberia sharply differ from each other in the character of the underlying surface. In western Siberia with its vast low-lying expanses, cloudiness rarely occurs because of convective movement of air-masses. Poorly defined topography and the cooling effect of large boggy areas hamper the development of thermal convection. During summer, stratal forms of clouds predominate here. Regions in the east adjoining the Ural mountains, where clouds of orographic origin predominate, are exceptions. In topographical character, eastern Siberia is entirely different from western Siberia. It consists of a combination of mountain ranges and plateaus, between which are situated undulating mountainous plateaus or low-lying areas with a developed hydrographic network. With the onset of summer, a vast low gradient region of relatively low pressure predominates over this territory, intensifying cyclonic activity which, in its turn, promotes the formation of convective forms of clouds.

However, formation of clouds suitable for impingement is irregular over eastern Siberian territory. In the northern regions adjoining the shores of the Arctic Ocean, stratal clouds, capable of causing drizzles, predominate during the warm period. In proportion to the distance from the sea to the interior of the land mass, there is less recurrence of stratal clouds, but, on the other hand, the number of days with convective clouds increases. The maximum number is observed in the zone adjoining 56°N, while a sharp fall is observed in the southern steppe regions, (with the exception of the Minusinsk and Tuvinsk basins), where the recurrence of days with convective clouds again increases. This trend is graphically presented in a qualitative manner in Fig. 44.

Formation of thick cumulus clouds over forest areas is explained by intense physical evaporation of rain water from the earth's surface and extensive transpiration from all stories of vegetation—maximum evaporation and transpiration occurring in the afternoon (13–15 hours). During the same period, the maximum development of cumulus clouds is observed, which indicates a close relationship between these processes.

As shown by N.G. Nesterov (1933), the barometric gradient falls more sharply above forest areas as compared to open spaces (plowed fields, felling areas, burnt-out areas, etc.). As a result of this, the heated air above the open spaces has anabatic currents, while on the other hand the more humid and cold air above forests has katabatic currents. As a result, colder and more humid air from the column above a forest, rises from the ground and moves toward the warmer air over the open space; and the reverse process is observed from the top. As a result of such circulation, the heated moist air rises up to the condensation level (1.2–1.5 km), where cumulus clouds are formed. This general pattern can be supplemented by vertical convection due to the rise of air masses on encountering slopes of mountain ranges and elevations, and also due to irregular heating of these slopes.

The sun heats the southern and southwestern slopes much more intensely as compared to northern and eastern slopes, the difference is their surface temperatures sometimes reaching tens of degrees. On coming in contact with the intensely heated surface of southern slopes, air is rapidly heated and rises upward, forming a stable anabatic current. This explains the frequent confinement of thick cumuli to the tops of mountain elevations.

In summer, the surfaces of large rivers and lakes have a lower temperatures during daytime as compared to surrounding areas. As a result of this, katabatic air currents can be observed above water reservoirs breaking up convective clouds.

In nature, a meteorological situation can sometimes be observed when all cloud formation factors are present, and yet formation of thick cumuli does not take place. Such a phenomenon is explained by the presence of a so-called inversion layer in the substandard atmospheric layers, in which the temperature rises, instead of dropping with altitude (as usually occurs in the atmosphere). In this case, even the large quantity of heat liberated during condensation of the water vapor in the air, while reaching the condensation level, is not sufficient to ensure a further rise of the mass of moist air in the warm inversion layer. In this case, the growth of clouds stops, or if the inversion layer is below the level of water vapor condensation, clouds do not form at all.

Development of a thick cumulus in the presence of an inversion layer is possible when a powerful heat flux (for example, from a large forest fire) pierces this layer. However, such clouds, in spite of their impressive size, often have a small water content and, on being planted with crystallizing reagents, yield a rainfall of short duration and low intensity.

Observations show that the formation of convective clouds is also hampered by screening of the atmosphere with smoke plumes from large forest fires stretching over several tens of kilometers, although beyond the limits of the smoke-filled zone, the cloud formation process is intensive.

Information on resource clouds without reference to forest fires has been given above. However, the data on prospective cloud cover during the period that large forest fires occur are of considerable interest for estimating the efficiency of the method of extinguishing forest fires with artificially induced rainfall. The concept of a 'large forest fire' has not yet acquired a quantitative definition. In isolated regions of Siberia and the Soviet Far East, forest fires engulf areas from a few hundred to tens of thousands of hectares. In the author's investigations, a large fire was defined as any fire in an area of 200 hectares or more. For western regions of the country, these values would be different. Therefore, it is possible to change the accepted quantitative assumptions in the process of conducting investigations.

In 1969, the duration of action of large forest fires on the Angara River bank territory was on the average 18 days, and two fires were active for over 30 days. During this period, rainfall was induced above these fires 12 times, i.e. every third day on the average. However, the length of the edge of these fires was so great that the area of rainfall precipitated from one cloud covered only a part of the fire edge.

Suppression of large forest fires with artificially induced rainfall in areas covered by the forest air protection service is carried out by special aerial units. The staff of a unit includes the crew of the airplane, the air observer and the flight aerologist. Considering that impact on clouds is carried out at comparatively high altitudes (about five-to-six km) twin-engined planes, *LI-2* and *IL-14* are used. A shot is fired into the cloud through a special vent in the fuselage of the plane. The planes are equipped with all navigation instruments ensuring normal working at high altitudes. If planes with hermetically sealed cockpits are used (for example, the *AN-24*), they are equipped with a special cassette with remote control from the cockpit for the discharges.

Artificially induced rainfall cannot always completely extinguish a forest fire, especially if the flame has penetrated into the forest floor or peat. Even in ground fires, old stumps, anthills, deadwood and hollows of trees can smolder for tens of days, creating a threat of revival. Therefore, the complete liquidation of the remaining sources and observation are essential operations. Usually, fire watching continues until frontal

Fig. 45. Infrared instruments for aerial detection of forest fires.

1—block of optical heads with scanning mirror with direct recording on film; *2*—light-sound fire detection indicator; *3*—recording system with recording of image and its development on paper; *4*—junction box; *5*—control panel; *6*—power supply unit.

precipitations (rainfall) occur which fully liquidate forest inflammability. This operation diverts a large number of workers during the most intensive period of fire risk.

For discovering hidden areas of combustion along the perimeter of the fire, airborne infrared apparatus, sensitive to heat radiation from forest landscape elements within the range of wavelengths from 1.8 to 14 microns, are used. Here there are two windows for transmission of infrared radiation between 1.8–5.3 and 8.0–14 microns. A photoelectric detector, which registers infrared radiation in a narrow solid angle, is used as the primary control (sensitive) element in the apparatus. Longitudinal scanning of the thermal image of the terrain is effected due to the progressive motion of the plane, while transverse scanning is effected by rotation of the receiver mirror, which runs past or scans the terrain within the range of a definite viewing angle (60–120°). The thermal image transformed into a visible image is photographed on 'running' photographic film, the speed of which is synchronized with the speed of the plane.

Figure 45 gives a general view of the aerial infrared apparatus, 'Fire Mapper System', used by the Canadian Forest Protection Service for detecting forest fires (Fire Mapper—Iris System, 1973). A block of optical heads with a scanning mirror is suspended under the fuselage of the patrolling plane, while the remaining blocks are arranged in the cockpit at places convenient for the operator. The viewing angle of the apparatus is 120°, the sensitivity of the thermal radiation receiver is 1°C with a background temperature of 20°C. The infrared image of the terrain is recorded by a light beam on an ordinary panchromatic film 70 mm in width. The drawing

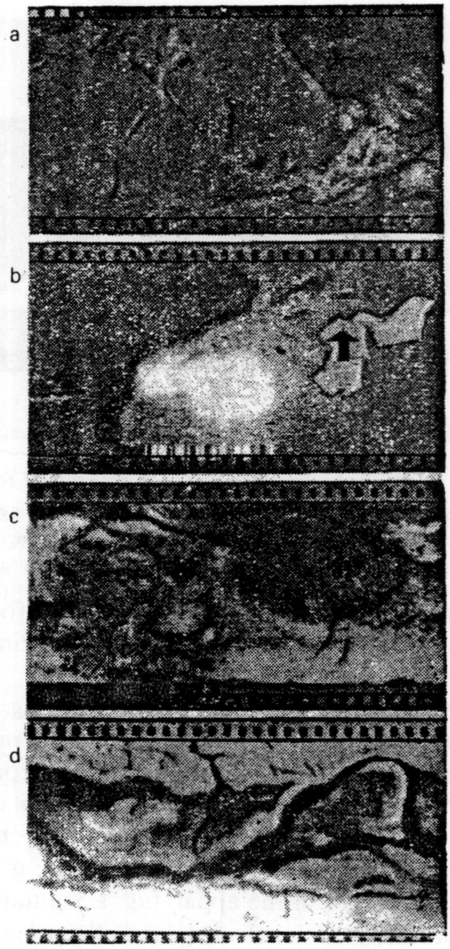

Fig. 46. Infrared image obtained by means of the airborne apparatus, 'Fire Mapper System'. *a*—areas of incipient fires among coniferous stands; *b*—image of a forest obtained by surveying through smoke clouds; *c*—unextinguished areas in freshly burnt clearings; *d*—flood plain of a river with soils of different humidity.

mechanism of the film is switched on when one of the lamps ignites on the indicator, showing the sector of the viewing angle where a hot spot has appeared.

Among infrared instruments the 'Fire Mapper System', is used not only for detecting hidden areas of combustion, but with its application it is also possible to carry out mapping of the perimeters of forest fires through smoke clouds, to chart areas with different soil humidities, to detect areas of forest stands damaged by pests and diseases of trees, etc. (Fig. 46). A distinctive characteristic of the 'Fire Mapper System' is the presence of the aim discriminator, which helps in distinguishing true and false aims in the picture. Light dashes at the edge of the picture obtained by means of this discriminator show that the light spots in the picture correspond to abnormally heated areas. The sensitivity of the receiver helps in reliable detection of 'hidden' fire areas within a space of 30×30 cm from an altitude of 600 m.

The use of infrared equipment affords prospects of detecting forest fires from high altitude planes and artificial earth satellites.

Bibliography

Alekseev, N.V. et al. 1971. Osnovy pozharnoi bezopasnosti (Principles of Fire Safety). Vysshaya shkola, Moscow, p. 245.

Amosov, G.A. 1958. Nekotorye osobennosti goreniya pri lesnykh pozharakh (Some Peculiarities of Combustion in Forest Fires). Leningrad, p. 29. Leningrad, p. 29.

Amosov, G.A. and N.N. Krasavina. 1958. Kharakter goreniya v lesu i primenenie ognetushitel'nykh smesei E's-1 i E's-2 (The Nature of Combustion in Forests and the Use of Fire-Extinguishing Mixtures *E'S-1* and *E'S-2*). *Sbornik rabot po lesnomu khozyaistvu.* Lesnaya promyshlennost', Moscow, p. 137–146.

Antsyshkin, S.P. 1957. Protivopozharnaya okhrana lesa (Forest Fire Protection). Goslesbumizdat, Moscow-Leningrad, p. 185.

Antsyshkin, S.P. and E.S. Artsybashev. 1967. Okhrana lesov ot pozharov (Protection of Forests from Fires). *Lesnoe khozyaistvo,* No. 10, p. 56–62.

Artsybashev, E.S. 1971. Infrakrasnaya ae'ros'emka lesnykh pozharov s vysotnykh samoletov i iskustvennykh sputnikov zemli (Infrared Aerial Survey of Forest Fires from High-Altitude Aircraft and Artificial Earth Satellites). *Lesnoe khozyaistvo,* No. 5, p. 60–64.

Artsybashev, E.S. and P.A. Gubin. 1970. Tushenie lesnykh pozharov iskusstvenno vyzyvaemymi osadkami iz oblakov (Extinguishing Forest Fires with Rainfall Induced Artificially from Clouds). *Lesnoe khozyaistvo,* No. 3, p. 73–76.

Artsybashev, E.S. et al. 1972. Vodoslivnoe oborudovanie k vertoletu KA-26 dlya tusheniya lesnykh pozharov s vozdukha (Spillway Apparatus in *KA-26* Helicopters for Extinguishing Forest Fires). *Lesnoe khozyaistvo,* No. 6, p. 61–62.

Artsybashev, E.S. and V.G. Lorberbaum. 1972. Khimicheskie metody bor'by s lesnymi pozharami (Chemical Methods of Controlling Forest Fires). Moscow, p. 46.

Artsybashev, E.S., V.P. Molchanov and O.K. Orlov. 1969. Zvukousilitel'naya ustanovka dlya vedeniya protivopozharnoi propagandy s patrul'nykh samoletov i vertoletov (Sound Amplifier for Conducting Antifire Messages from Patrol Aircraft and Helicopters). *Lesnoe khozyaistvo,* No. 11, p. 54–57.

Artsybashev, E.S. and O.K. Orlov. 1968. Televizionnaya ustanovka dlya obnaruzheniya lesnykh pozharov (Television Device for Detecting Forest Fires). *Sbornik statei po itogam dogovornykh nauchno-issledovatel'skikh rabot za 1965–1966 gg.* Lesnaya promyshlennost', Moscow, p. 142–148.

Artsybashev, E.S. and B.G. Shtuchkov. 1965. Nazemnye sredstva i sposoby obnaruzheniya lesnykh pozharov (Ground Aids and Methods of Detecting Forest Fires). In *Sovremennye Voprosy Okhrany Lesov ot Pozharov i Bor'by s Nimi,* Moscow, p. 119–133.

Artsybashev, E.S. and L.V. Stolyarchuk. 1971. Analiz resursnoi oblachnosti, perspektivnoi dlya vyzyvaniya osadkov nad lesnymi pozharami (Analysis of Cloud Cover Resources with a Possibility of Inducing Rainfall over Forest Fires). *Lesnoe khozyaistvo,* No. 9, p. 57–60.

Australian Forest Industries Journal, vol. 39, No. 3, 1973, p. 31.

Back Pack Pumps. Wajax Fire control equipment, Montreal, 1972, p. 41.

Balbyshev, I.N. 1949. Priroda pozharov zapadnosibirskoi taigi (Nature of Fires in the West Siberian Taiga). *Lesnoe khozyaistvo,* No. 8, p. 51–54.

Barrows, J.S. 1960. Control of Lightning Fires in American Forests (Fifth World Forestry Congress). Washington, Vol. 2, p. 851–856.

Belov, S.V., I.D. Dmitriev and A.E. Kolosova. 1962. Aérofots'emka i aviatsiya v lesnom khozyaistve (Aerial Photography and Aviation in Forestry). Leningrad, p. 255.

Biryulin, V.I., K.N. Makarov and A.N. Kanishchev. 1969. Vertolety v narodnom khozyaistve (Helicopters in National Service). Transport, Moscow, p. 175.

Bugai, B.K., S.V. Gundar, F.F. Mishkov, A.I. Pakhomov and G.P. Telitsyn. 1972. Ruchnye orudiya i rantsevaya apparatura dlya tusheniya lesnykh pozharov (Hand-Operated Implements and Haversack Equipment for Extinguishing Forest Fires). TsBTI Gosleskhoza SSSR, Moscow, p. 23.

Chelyadinova, A.I. 1945. Kolichestvo khvoi i kharakter ee razvitiya u sosny (Number of Needles in Pines and the Nature of Their Development). *Doklady Vsesoyuznogo Soveshchaniya po Fiziologii Rastenii,* vol. 2, p. 30–50.

Chervonnyi, M.G. 1973. Ser'eznoe vnimanie protivopozharnoi profilaktike (Detailed Study of Fire-Prevention Measures). *Lesnoe khozyaistvo,* No. 3, p. 46–48.

Dolgopolov, V.G. and G.M. Kulakov. 1972. Primenenie sovremennoi telemetricheskoi tekhniki v lesnom khozyaistve (Use of Modern Telemetric Technique in Forestry). TsBNTI Gosleskhoza SSSR, Moscow, p. 83.

Domanitskii, A.P., R.G. Dubrovina and A.I. Isaeva. 1971. Reki i ozera Sovetskogo Soyuza (Rivers and Lakes of the Soviet Union). Gidrometeoizdat, Leningrad, p. 101.

Egorov, N.M. 1955. O nekotorykh detalyakh puska vstrechnogo nizovogo ognya (Some Details of Starting a Backfire). *Lesnoe khozyaistvo,* No. 9, p. 55–57.

Elpat'evskii, M.M., M.P. Elpat'evskii and V.K. Konstantinov. 1970. Osushenie i osvoenie zabolochennykh lesnykh zemel' (Reclamation and Development of Swamped Forest Soils). Lesnaya promyshlennost', Moscow, p. 232.

Ershov, E.V. 1962. Instruktivnye ukazaniya po tekhnicheskomu obsluzhivaniyu i e'kspluatatsii zazhigatel'nogo apparata ZA-1M (Instructions on Technical Service and Operation of Ignition Apparatus, *ZA-1M*). Leningrad Scientific Research Institute of Forestry, Leningrad, p. 10.

Firemapper—ires system. Computing devices of Canada Limited. Vancouver, 1973, p. 22.

Forester, D. Fire Chemicals. British Columbia Lumberman, vol. 52, No. 7, p. 37–39.

Geiger, R. 1960. Klimat prizemnogo sloya vozdukha (Climate of Surface Atmospheric Layer). Inostrannaya literatura, Moscow, p. 486.

Golden Arrow Tanker. Wajax fire control equipment. Montreal, 1972, p. 47.

Gorbatova, N.G. 1963. Teplotvornaya sposobnost' nekotorykh raznovidnostei lesnykh goryuchikh materialov (Calorific Values of Some Types of Forest Combustibles). In *Lesnye Pozhary i Bor'ba s Nimi.* Moscow, p. 158–162.

Hastings, J. 1956. Television Testing for Forest Fire Detection in California. Fire Control Notes, No. 1, p. 24–26.

Knorre, G.F. 1959. Chto takoe gorenie? (What is Combustion?). E'nergoizdat, Moscow, p. 119.

Konstantinov, V.K. and I.A. Yuzepchuk. 1972. Nekotorye voprosy osusheniya bolot s bednymi torfami (Some Problems of Reclamation of Boggy Soils with Poor Turf). In *Issledovaniya po Lesnomu Khozyaistvu.* Leningrad, p. 317–331.

Korchagin, A.A. 1954. Usloviya vozniknoveniya pozharov i gorimost' lesov evropeiskogo severa (Causes of Fire Origin and Combustibility of North European Forests). *Uchenye zap. LGU, Ser. geograf. nauk.,* vol. 8, p. 182–322.

Korovin, G.N. and N.S. Loginova. 1971. Raschet optimal'nogo rasshireniya patrul'nykh poletov (Computation of Optimum Increase in Patrol Flights). *Sbornik programm obrabotki lesokhozyaistvennoi informatsii na E'VM 'Minsk-22',* Leningrad Scientific Research Institute of Forestry, Leningrad, p. 45–64.

Korovin, G.N. 1967. Osobennost' rascheta perimetrov lesnykh nizovykh pozharov (Special Features of Calculating Perimeters of Forest Ground Fires). *Sbornik nauchno-issledovatel'skikh rabot po lesnomu khozyaistvu,* vol. 11, p. 330–345.

Krasavina, N.N. 1965. Ognezashchitnye i ognegasyashchie svoistva vodnykh rastvorov neorganicheskikh veshchestv v bor'be s lesnymi pozharami (Fire Resistant and Fire-Extinguishing Properties of Aqueous Solutions of Inorganic Substances in Forest Fire Control). In *Sovremennye Voprosy Okhrany Lesov ot Pozharov i Bor'ba s Nimi.* Moscow, p. 134–159.

Krasavina, N.N. 1963. Tekhnicheskie ukazaniya po primeneniyu khimicheskikh veshchestv na tushenii lesnykh pozharov (Technical Instructions on the Use of Chemicals for Extinguishing Forest Fires). Leningrad Scientific Research Institute of Forestry, Leningrad, p. 17.

Kurbatskii, N.P. 1970. Issledovanie kolichestva i svoistv lesnykh goryuchikh materialov (Research on Quantity and Properties of Forest Combustibles). In *Voprosy Lesnoi Pirologii,* Krasnoyarsk, p. 5–58.

Kurbatskii, N.P. 1970. O klassifikatsii lesnykh pozharov (Classification of Forest Fires). *Lesnoe khozyaistvo,* No. 3, p. 68–73.

Kurbatskii, N.P. 1962. Tekhnika i taktika tusheniya lesnykh pozharov (Technique and Strategy of Extinguishing Forest Fires). Goslesbumizdat, Moscow, p. 153.

Kurbatskii, N.P. 1958. Tushenie lesnykh pozharov vodoi (Extinguishing Forest Fires with Water). Leningrad Scientific Research Institute of Forestry, Leningrad, p. 58.

Kurbatskii, N.P. 1972. Vidy otzhiga i ikh primenenie dlya lokalizatsii lesnykh pozharov (Types of Annealing and their Use in Localizing Forest Fires). In *Voprosy Lesnoi Pirologii,* Krasnoyarsk, p. 153–163.

Kutsenogii, K.P. 1970. K vozmozhnosti ispol'zovaniya vodyanogo ae'rozolya dlya bor'by s lesnymi pozharami (Possibility of Using Aqueous Aerosol for Controlling Forest Fires). In *Voprosy Lesnoi Pirologii,* Krasnoyarsk, p. 340–352.

Khalevitskii, Z.Z. 1971. Meteorologiya na sluzhby okhrany lesov (Meteorology in the Service of Forest Protection). Gidrometeoizdat, Moscow, p. 88.

Lange, S. 1963. Sechs Jahrzehnte Feuerwachtürme. *Archiv für Forstwesen,* vol. 12, No. 6, p. 28–34.

Len'kova, A. 1971. Oskal'pirovannaya zemlya (Bare Land). Translated from the Polish. Progress, Moscow, p. 285.

Lockman, M.R. 1970. Forest Fire Losses in Canada. Ottawa, p. 13.

Lorberbaum, V.G. and N.V. Bashun. 1965. Tekhnicheskie ukazaniya po tusheniyu lesnykh, torfyanykh i podstilochno-gumusovykh pozharov rastvorami poverkhnostno-áktivnykh veshchestv (Technical Instructions on Extinguishing Forest Peat and Litter Humus Fires with Solutions of Surface-Active Substances). Leningrad Scientific Research Institute of Forestry, Leningrad, p. 15.

Macleod, I.C. 1960. Research in Forest Fire (Fifth World Forestry Congress). Washington, p. 627–632.

Melekhov, I.S. 1939. Opyt izucheniya lesnykh pozharov v lesakh Severa (Study of Forest Fires in Northern Forests). AITI, Arkhangel'sk, p. 39.

Melekhov, I.S. 1947. Priroda lesa i lesnye pozhary (Nature of Forest and Forest Fires). Arkhangel'sk, p. 57.

Melekhov, I.S. 1948. Vliyanie pozharov na les (Effect of Fires on Forests). Goslestekhizdat, Moscow-Leningrad, p. 124.

Mokeev, G.A. and V.M. Speranskii. 1965. Instruktsiya po e'kspluatatsii rantsevogo lesnogo opryskivatelya RLO (Operational Instructions for the Forest Haversack Sprayer *RLO*). Leningrad Scientific Research Institute of Forestry, Leningrad, p. 11.

Molchanov, A.A. 1960. Gidrologicheskaya rol'lesa (Hydrological Role of Forest). Izd-vo AN SSSR, Moscow, p. 484.

Molchanov, A.A. 1961. Les i klimat (Forest and Climate). Izd-vo AN SSSR, Moscow, p. 279.

Molchanov, A.A. 1954. Vliyanie lesnykh pozharov na drevostoi (Effect of Forest Fires on Forest Stands). *Trudy Instituta lesa AN SSSR,* vol. 16, p. 314–335.

Molchanov, A.A. 1949. Zapasy khvoi v sosnovykh drevostoyakh razlichnogo vozrasta (Needle Contents in Pine Stands of Various Ages). *Doklady AN SSSR,* vol. 67, No. 5, p. 909–912.

Molchanov, V.P. and I.A. Proshkin. 1956. Vertolet MI-4 na bor'be s lesnymi pozharami (Helicopter *MI-4* in Forest Fire Control). Leningrad Scientific Research Institute of Forestry, Leningrad, p. 15.

Molchanov, V.P. 1957. Ob izmenenii vlazhnosti khvoi sosny obyknovennoi v svyazi s verkhovymi pozharami (Changes in Needle Humidity in Scotch Pines due to Tree-Crown Fires). *Botanicheskii zhurnal,* No. 2, p. 46–51.

Molchanov, V.P. 1965. Opredelenie zapasov goryuchikh materialov v pologe sosnovykh nasazhdenii (Determination of Fuel Contents in the Canopy of Pine Forest Stands). In *Sovremennye Voprosy Okhrany Lesov ot Pozharov i Bor'by s Nimi.* Moscow, p. 108–118.

Molchanov, V.P. 1957. Usloviya rasprostraneniya verkhovykh pozharov v sosnyakakh (Causes of the Spread of Tree-Crown Fires in Pine Forests). *Lesnoe khozyaistvo,* No. 8, p. 50–53.

Molchanov, V.P. 1961. Vertikal'noe raspredelenie skorosti vetra v sosnovykh nasazhdeniyakh (Vertical Distribution of Wind Velocity in Pine Stands). *Sbornik rabot po lesnomu khozyaistvu,* Leningrad Scientific Research Institute of Forestry, vol. 4, p. 292–301.

Nesterov, V.G. 1949. Gorimost' lesa i metody ee opredeleniya (Combustibility of Forests and Methods of its Determination). Goslesbumizdat, Moscow, p. 74.

Nesterov, V.G. 1945. Pozharnaya okhrana lesa (Protection of Forests from Fires). Goslestekhizdat, Moscow, p. 170.

Nesterov, N.S. 1933. Ocherki po lesovedeniya (Articles on Silviculture). Goslestekhizdat, Moscow, p. 247.

New Scientist, vol. 36, No. 438, London, 1967, p. 79.

Pozdnyakov, L.K. et al. 1969. Biologicheskaya produktivnost' lesov Srednei Sibiri i Yakutii (Biological Productivity of the Central Siberian and Yakutskii Forests). Krasnoyarsk, p. 155.

Prikhot'ko, G.F. 1968. Iskusstvennye osadki iz konvektivnykh oblakov (Artificial Rainfall from Convective Clouds). Gidrometeoizdat, Leningrad, p. 134.

Rode, A.A. 1955. Pochvovedenie (Soil Science). Goslesbumizdat, Moscow-Leningrad, p. 524.

Rutkovskii, V.I. 1936. Issledovanie zaderzhaniya osadkov travyanym i mokhovym pokrovom (Research in Absorption of Rainfall by Peat and Moss Cover). *Meteorologiya i gidrologiya,* No. 11.

Savina, A.V. 1941. Izuchenie vliyaniya rubok ukhoda na svetovoi rezhim (Study of the Effect of Felling on Light Conditions). *Trudy VNIILKh,* vol. 21, p. 5–29.

Semenov, N.N. 1957. Tsepnye reaktsii (Chain reactions). *BE'S,* vol. 46, Moscow, p. 562-567.

Serebrennikov, P.P. and V.V. Matreninskii. 1937. Lesnye pozhary i bor'ba s nimi (Forest Fires and their Control). Goslestekhizdat, Moscow, p. 177.

Shtuchkov, B.G. 1965. Instruktsiya po obsluzhivaniyu i e'kspluatatsii pozharnoi nablyudatel'noi machty PNM-3 (Instructions on Servicing and Operating the Fire-Observation Mast *PNM-3.* Leningrad Scientific Research Institute of Forestry, Leningrad, p. 13.

Smirnov, V.V. 1961. Izmenenie okhvoeniya i oblistveniya v elovykh i elovo-listvennykh drevostoyakh srednei taigi v svyazi s ikh vozrastom (Changes in the Development of Needles and Foliage in Spruce and Broad-Leaved Forest Stands, Respectively, of the Central Taiga in Relation to their Age). *Trudy laboratorii lesovedeniya,* vol. 3, p. 235–263.

Snyktin, G.V. 1969. Zapasy goryuchikh materialov nekotorykh tipov listvennichnykh lesov i zaroslei kedrovogo stlanika v basseine r. Kolymy (Fuel Content of Some Types of Broad-Leaf Forests and Thickets of Dwarf Stone Pine in the Basin of the Kolyma River). *Sbornik trudov Dal'NIILKh,* No. 9, p. 443–446.

Sofronov, M.A. 1967. Lesnye pozhary v gorakh Yuzhnoi Sibiri (Forest Fires in the Mountains of Southern Siberia). Nauka, Moscow, p. 149.

Sofronov, M.A. 1970. Ob uslov i yakh vysykhaniya lesnykh goryuchikh Materialov pod pologom drevostoev (Drying Conditions of Forest Combustibles under Forest Stand Canopies). In *Voprosy Lesnoi Pirologii,* Krasnoyarsk, p. 59–103.

Speranskii, V.M. and B.G. Shtuchkov. 1967. Instruktsiya po ékspluatatsii lesopozharnoi s'emnoi avtotsisterny TsOS (Instructions on the Operation of the Forest Fire Detachable Tank-Truck, *TsOS*). Leningrad Scientific Research Institute of Forestry, Leningrad, p. 13.

Speranskii, V.M. 1970. Instruktivnye ukazaniya po obsluzhivaniyu i e'kspluatatsii malogabaritnoi lesnoi perenosnoi motopompy ML-100 (Instructions on Servicing and Operation of the Small Portable Motor-pump *ML-100*) (MLAZ). Leningrad Scientific Research Institute of Forestry, Leningrad, p. 15.

Speranskii, V.M. 1964. Instruktivnye ukazaniya po tekhnicheskomu obsluzhivaniyu i e'kspluatatsii rantsevogo pnevmaticheskogo ognetushitelyaopryskivatelya ROOP-4A (Instructions on the Technical Servicing and Operating the Fire-Observation Mast *PNM-3*). Leningrad Scientific Sprayer *ROOP-4A*). Leningrad Scientific Research Institute of Forestry, Leningrad, p. 11.

Speranskii, V.M. 1961. Instruktsiya po obsluzhivaniyu i e'kspluatatsii pozharnoi nablyudatel'noi machty PNM-2 (Instructions on Servicing and Operation of the Fire Observation Mast *PNM-2*). Leningrad Scientific Research Institute of Forestry, Leningrad, p. 15.

Speranskii, V.M. 1955. Rantsevyi Diafragmovyi opryskivatel' RDOS-1 (Ustroistvo, Primenenie, Ukhod) (Haversack Diaphragm Sprayer *RDOS-1* (Design, Application and Maintenance). Leningrad Scientific Research Institute of Forestry, Leningrad. p. 10.

Steven, S. 1963. Michigan Sand Caster—Model 111. *Journ. Fire Control Notes,* vol. 24, No. 4, p. 95–97.

Tamarkin, M.L. 1966. Okhrana lesov ot pozharov v Severnoi Amerike (Forest Fire Protection in North America). Lesnaya promyshlennost', Moscow, p. 135.

Telitsyn, G.P. and A.P. Sosnovshchenko. 1969. Teplotvornaya sposobnost' nekotorykh lesnykh goryuchikh materialov na Dal'nem Vostoke (Calorific Value of Some Combustibles in the Far East). *Sbornik trudov Dal'NIILKh,* p. 439–441.

Telitsyn, G.P. 1965. Zavisimost' skorosti rasprostraneniya nizovykh pozharov ot uslovii pogody (Relationship of Ground Fire Spread with Weather Conditions). *Sbornik trudov Dal'NIILKh,* vol. 7, p. 391–405.

The New Zealand Timber Journal, vol. 17, No. 10, 1971, p. 35.

The Timberman, No. 6, 1961, Iourk.

The Timberman, vol. 56, No. 9, 1955, p. 80.

Tkachenko, M.E. 1952. Obshchee lesovodstvo (General Forestry). Goslesbumizdat, Moscow-Leningrad, p. 599.

Uspenskii, S.N. 1959. Zagoranie lesa ot molnii i mery preduprezhdeniya pozharov v lentochnykh borakh Priirtysh'ya (Forest Fires due to Lightning and Fire Prevention Measures in Forest Belts near the Irtish

Mountains). *Trudy kazakhskogo NII lesnogo khozyaistva*, vol. 2.

Valdaiskii, N.P. and A.N. Chukichev. 1969. Ruchnoi motorizovannyi gruntomet (Instruktivnye ukazaniya po montazhu i e'kspluatatsii) (Portable Motorized Sand Caster (Instructions on Mounting and Operation)). Leningrad Scientific Research Institute of Forestry, Leningrad, p. 15.

Valdaiskii, N.P. 1970. Issledovaniya po sozdaniyu gruntometov dlya tusheniya lesnykh nizovykh pozharov (Research on Development of Sand Casters for Extinguishing Forest Fires). *Sbornik materialov nauchnoi konferentsii po voprosam lesnogo khozyaistva,* Pushkino, p. 28–30 (VNIILM).

Valdaiskii, N.P., Yu.M. Kodyakov and A.N. Chukichev. 1972. Frezernyi polosoprokladyvatel' PF-1 (Milling Strip Plier *PF-1*). Leningrad Scientific Research Institute of Forestry, Leningrad, p. 19.

Valendik, E'.N. 1968. Veter i lesnoi pozhar (Wind and Forest Fire). Nauka, Moscow, p. 116.

Vanin, S.I. 1934. Drevesinovedenie (Timber Science). Goslestekhizdat, Leningrad, p. 548.

Vonskii, S.M. 1957. Intensivnost' ognya nizovykh lesnykh pozharov i ee prakticheskoe znachenie (Practical Significance of the Flame Intensity of Forest Ground Fires). Izd. Leningrad Scientific Research Institute of Forestry, No. 52, p. 108–117.

Vonskii, S.M. and V.A. Zhdanko. 1969. Metodicheskie ukazaniya po sostavleniyu i primeneniyu mestnykh shkal pozharnoi opasnosti v lesu (Methodological Instructions on Organization and Use of Localized Control of Forest Fires). Leningrad Scientific Research Institute of Forestry, Leningrad, p. 21.

Vonskii, S.M. 1971. Metodika rascheta chislennosti rabochikh, vooruzhennykh rantsevoi apparaturoi dlya tusheniya lesnykh nizovykh pozharov (Method of Estimating the Personnel Strength of Workers Equipped with Haversack Apparatus for Extinguishing Forest Fires). Leningrad Scientific Research Institute of Forestry, Leningrad, p. 26.

Wajax *Mark-3* Centrifugal. Wajax Fire Control Equipment, Montreal, 1972, p. 3–3a.

Wally, T. 1971. The Big Burn. *Light Mag.*, vol. 60, No. 5, p. 14–16.

Water Handling Equipment Guide. US Department of Agriculture Forest Service, November, 1969, p. 113.

Zamyslovskii, V.D. 1961. Instruktivnye ukazaniya po tekhnicheskomu obsluzhivaniyu i e'kspluatatsii zazhigatel'nogo apparat ZA-1 (Instruction Manual on Technical Servicing and Handling of Fire Equipment *ZA-1*). Leningrad Scientific Research Institute of Forestry, Leningrad, 13 p.

Zhdanko, V.A. 1965. Nauchnye osnovy postroeniya mestnykh shkal i znachenie ikh pri razrabotke protivopozharnykh meropriyatii (Scientific

Principles of Constructing Local Scales and their Significance in Devising Antifire Measures). In *Sovremennye Voprosy Okhrany Lesov ot Pozharov i Bor'by s Nimi,* Moscow, p. 53–85.

Zhukovskaya, V.I. Uvlazhenie i vysykhanie gidroskopicheskikh lesnykh goryuchikh materialov (Moistening and Drying of Hygroscopic Forest Combustibles). In *Voprosy Lesnoi Pirologii,* Krasnoyarsk, p. 105–141.